110 Topics in Current Chemistry

Fortschritte der Chemischen Forschung

Managing Editor: F. L. Boschke

Modern Syntheses
of Cobalt (III) Complexes

By M. Shibata

Guest Editor: H. Yamatera

With 44 Figures and 31 Tables

 Springer-Verlag Berlin Heidelberg GmbH 1983

Professor Dr. Muraji Shibata †
Kanazawa University
Faculty of Science
Department of Chemistry
I-I Marunouchi
Kanazawa, Ishikawa, 920, Japan

This series presents critical reviews of the present position and future trends in modern chemical research. It is addressed to all research and industrial chemists who wish to keep abreast of advances in their subject.

As a rule, contributions are specially commissioned. The editors and publishers will, however, always be pleased to receive suggestions and supplementary information. Papers are accepted for "Topics in Current Chemistry" in English.

ISBN 978-3-662-15306-2   ISBN 978-3-540-39486-0 (eBook)
DOI 10.1007/978-3-540-39486-0

Library of Congress Cataloging in Publication Data.
Shibata, Muraji, 1924 — Modern syntheses of cobalt(III) complexes.
(Topics in current chemistry; 110) Bibliography: p. Includes index.
1. Cobalt compounds — Synthesis. 2. Complex compounds — Synthesis.
I. Title. II. Series.
QD1.F58 vol. 110   540s [546'.6232]   82-19361   [QD181.C6]

Managing Editor:

Dr. *Friedrich L. Boschke*
Springer-Verlag, Postfach 105280, D-6900 Heidelberg 1

Guest Editor of this volume:

Prof. Dr. *Hideo Yamatera*, Nagoya University, Department of
Chemistry, Faculty of Science, Furo-cho, Chikusa-cho,
Nagoya, Japan, 464

Editorial Board:

Preface

The developments in the studies based on various spectroscopic techniques, X-ray studies, ligand field theory, conformation analysis, theory of optical activity, etc. have engendered a great interest in the stereochemistry of transition metal complexes. It should be emphasized, however, that a rather slow but steady progress in preparative studies of the metal complexes have exerted itself in the background of the recent advance in stereochemistry.

My entrance into the preparative studies on cobalt(III) complexes was in the early 1950's. In those days, the studies by means of visible and ultraviolet spectra and infrared spectra were very popular. But, most material on research was obtain able from the classical literatures in the Jørgensen-Werner period. The preparative method of cobalt(III) complexes was transformed, in the 1960's, by the development of some elegant methods and by the use of chromatographic techniques in the preparative methods.

The "tricarbonato-method" in which tricarbonato-cobaltate(III) anion, $[Co(CO_3)_3]^{3-}$, was used as the starting material, has been developed mostly by the author and his coworkers. By this method, many new complexes have been synthesized, and the versatility of the method has been generally recognized.

In this monograph, the author gives some fundamental concepts in the latest methods using the "tricarbonato-method". This monograph has been organized as follows:
a) Each chapter begins with historical or general aspects.
b) For the preparations practical procedures are described.
c) Some topics of stereochemistry are added in order to illustrate trends in the research.
d) Figures and numerical data of electronic and circular dichroism spectra are cited.

I should like to thank Professor H. Yamatera of Nagoya University, who is the editor of this volume. I want to thank Dr. F. Boschke of Springer-Verlag. I should like to record my gratitude to my students and colleagues in the department of chemistry of Kanazawa University.

July 1982   Muraji Shibata

Professor Muraji Shibata, born in 1924, graduated from the Tokyo Institute of Technology and received his D. Sc. from the University of Tokyo. After having been an associate professor at Kanazawa University and then a professor at Ibaraki University, he held the chair of coordination chemistry at Kanazawa University since 1964 until he died September 15, 1982.

His ingenuity was displayed in the syntheses of coordination compounds. He established the 'tricarbonato method', a versatile method for the preparation of cobalt complexes. Using this and other ingeneous techniques, he prepared more than three hundred new complexes and carried out spectrochemical and stereochemical investigations of these new complexes.

He was not only a creative scientist, but also a stimulating teacher. His book "Introduction to Coordination Chemistry" is very popular among chemistry students in Japan.

H. Yamatera

Table of Contents

Abbreviations

acac	acetylacetonate
ala	alaninate
β-ala	β-alaninate
alama	alanine-N-monoacetate
aspn (or asn)	asparaginate
atc	3-acetylcamphorate
bn	1,3-butanediamine
bpy	2,2′-bipyridine
bzac	benzoylacetonate
chxn	1,2-cyclohexanediamine
cptn	1,2-cyclopentanediamine
dabp	2,2′-diaminobiphenyl
daes	di(2-aminoethyl)sulfide
dba	diaminobutyrate
dema	N-methylbis(2-aminoethyl)amine
	(4-methyldiethylenetriamine)
dien	diethylenetriamine
dmedds	N,N′-dimethylethylenediaminedisuccinate
don	1,12-dodecamethylenediamine
edda	ethylenediamine-N,N′-diacetate
eddams	ethylenediamine-N-N-diacetic-N′-mono-
	succinate
eddp	ethylenediamine-N,N′-di-L-α-propionate
edds	N,N′-ethylenediaminedisuccinate
edta	ethylenediaminetetraacetate
eee	1,8-diamino-3,6-dithiaoctane
en	ethylenediamine
ete	1,9-diamino-3,7-dithianonane
glu	glutamate
gly	glycinate
glygly (or Hgg)	glycylglycinate
gly-L-leu	glycyl-L-leucinate
Hasp	hydrogenaspartate
Heta	2-aminoethanol (ethanolamine)
Hhmcar	hydroxymethylenecarvone
Hhmpul	hydroxymethylenepulegone
hmc	hydroxymethylenecamphorate

Hpra	2-amino-1-propanol (propanolamine)
ibn	2-methyl-1,2-propanediamine
ida	iminodiacetate
ileu	isoleucinate
i-dtma	iso-diethylenetriaminemonoacetate
leu	leucinate
linpen	1,14-diamino-3,6,9,12-tetraazatetradecane
lys	lysinate
mal	malonate
malato	malate
medds	N-methylethylenediaminedisuccinate
mepenten	N,N,N′,N′-tetrakis(2′-aminoethyl)-1,2-propanediamine
metacn	2-methyl-1,4,7-triazacyclononane
met	methioninate
mida	methyliminodiacetate
ox	oxalate
pdc	pyridine-2,6-dicarboxylate
pdta	propylenediaminetetraacetate(1,2-propanediaminetetraacetate)
penten	N,N,N′,N′-tetrakis(2-aminoethyl)ethylene-diamine
pheala-gly	phenylalanylglycinate
phen	1,10-phenanthroline
pic	2-pyridinecarboxylate (picolinate)
pn	propylenediamine (1,2-propanediamine)
pro	prolinate
proma	proline-N-monoacetate
promp	prolin-N-monopropionate
ptn	2,4-pentanediamine
ptnta	2,4-pentanediaminetetraacetate
py	pyridine
rdas	rac-diaminosuccinate
sar	sarcosinate
ser	serinate
tacn	1,4,7-triazacyclononane
tame	1,1,1-tris(aminomethyl)ethane
tart	tartrate
tet	1,10-diamino-4,7-dithiadecane
2,3,2-tet	3,7-diaza-1,9-nonanediamine
3.2.3-tet	4,7-diaza-1,10-decanediamine
tfac	trifluoroacetylacetonate
thr	threoninate
tmd	1,4-butanediamine
tn	trimethylenediamine
trien	triethylenetetramine
TTP	1,4,8,11-tetrathiacyclotetradecane

| TTX | 13,14-benzo-1,4,8,11-tetrathiacyclopenta-decane |
| val | valinate |

1 Some Modern Methods of General Syntheses

1.1 Significance of Preparative Work

1.1.1 Around Triamminetrinitrocobalt(III)

Complexes with the general formula $[Co(NO_2)_n(NH_3)_{6-n}]$ (n = 1–5; omit the charges) are typical Werner-type complexes and most complexes of the series known up to the age of Werner. However, the series still remain incomplete up to now.

Triamminetrinitrocobalt(III), $[Co(NO_2)_3(NH_3)_3]$, was first prepared in 1866 by Erdmann, [1] later by Werner [2] and Jørgensen, [3] who prepared the complex by the air-oxidation of ammoniacal cobalt(II) salt solutions containing sodium nitrite and a large amount of ammonium chloride. In 1938, Duval [4] examined the products obtained from several different procedures by absorption and infrared spectroscopy, refractive index of aqueous solutions, conductivity, and X-ray powder diffraction. He recognized two products in the Werner's preparation and the Jørgensen's preparation. In that year, Sueda [5] reported an isomeric complex from the reaction of the $[Co(NO_3)_3(NH_3)_3]$ complex [6] with sodium nitrite in a cold aqueous solution, which was assumed to be *cis-cis* isomer on the basis of the absorption spectrum.

In 1952, the X-ray crystal analysis of the Jørgensen's product by Kuroya et al. [7] determined it to be a *trans-cis* isomer. Three years later, the IR spectral examination allowed Majumdar et al. [8] to conclude that Werner's preparation yielded the *trans-cis* isomer and Jørgensen's preparation the *cis-cis* isomer. In the same year Shibata et al. [9] reinvestigated Sueda's preparation by aquating the $[Co(NO_3)_3(NH_3)_3]$ complex in a cold aqueous solution containing a small amount of acetic acid. They suspected that the product was *cis-cis*-$[Co(NO_2)_3(NH_3)_3]$ on the basis of the IR peak at 1052 cm^{-1}, in view of the fact that the IR spectrum of $[Co(ONO)(NH_3)_5]Cl_2$ showed a Co-ONO characteristic peak at 1065 cm^{-1} [10].

In 1964, Maddock and Todesco [11] reported the IR spectral studies of a number of cobalt(III) complexes; in their report, they concluded that the *cis-cis* or *fac* isomer of Majumdar et al. [8], which still gave the nonelectrolyte composition, was in fact an ionisation isomer of the $[Co(NO_2)_2(NH_3)_4][Co(NO_2)_4(NH_3)_2]$ type.

In the same year, Shibata et al. [12] reported a general method for preparing the ammine-nitro series, starting with potassium tricarbonatocobaltate(III). In addition to established complexes, $[Co(NO_2)(NH_3)_5]^{2+}$, *cis*-$[Co(NO_2)_2(NH_3)_4]^+$, *mer(trans-cis)*-$[Co(NO_2)_3(NH_3)_3]$ and *trans*-$[Co(NO_2)_4(NH_3)_2]^-$, a new amminepentanitro complex, $[Co(NO_2)_5(NH_3)]^{2-}$, was prepared as potassium salts.

Because of the less clear situation with *fac(cis-cis)*-$[Co(NO_2)_3(NH_3)_3]$, Hagel and Druding [13] reexamined the products prepared by different methods by chromatography and electrophoresis. They stated that for the *fac* isomer the method described by Shibata et al. [12], who claimed to have produced the *mer* isomer, gave the best yield, while for the *mer* isomer Jørgensen's method [3] modified by Cooley et al. [14] was the best way of obtaining a pure isomer.

In 1978, Siebert [15] prepared the *fac* isomer as follows;

$$[Co(NH_3)_3(H_2O)_3]^{3+ \ [16]} \xrightarrow[\substack{H_2O \text{ with } CH_3COOH \\ (-5 \sim -8\,°C)}]{NO_2^-} fac\text{-}[Co(ONO)_3(NH_3)_3]$$

$$\xrightarrow[\substack{H_2O \\ (\text{room temp.})}]{} fac\text{-}[Co(ONO)(NO_2)_2(NH_3)_3] \xrightarrow[\substack{H_2O \\ (50\,°C)}]{} fac\text{-}[Co(NO_2)_3(NH_3)_3]$$

The three *fac* complexes were characterized mainly by IR, absorption and ^{1}H-NMR spectra. Siebert has clarified that the product of Shibata et al. [9] was evidently the intermediate *fac*-[Co(ONO)(NO_2)_2(NH_3)_3].

Thus, two isomers of triamminetrinitrocobalt(III) were finally recognized after more than hundred years of research. However, a complex, *cis*-[Co(NO_2)_4(NH_3)_2]$^-$, is still missing in this series, although the corresponding *trans* isomer is the very familiar Erdmann's salt, NH_4[Co(NO_2)_4(NH_3)_2] [17].

1.1.2 Reviews of the Common Methods

The usual methods for cobalt(III) complexes may be classified into two categories, the oxidation from Co(II) to Co(III), and the substitution between a starting cobalt(III) complex and a reagent for ligand. Some examples are:

1) Hexaamminecobalt(III) salts may be prepared by the oxidation of cobalt(II) ion in ammoniacal solution: i) Air oxidation that gives a pentaammine intermediate, which is then converted to the hexaammine by heating with aqueous ammonia under pressure; ii) Oxidation with an oxidizing agent such as hydrogen peroxide; iii) Oxidation with a catalyst establishing the equilibrium between the pentaammine and hexaammine ions at room temperature and atmospheric pressure [18]. The last and best method uses decolorizing charcoal as the catalyst.

$$4\,CoCl_2 + 4\,NH_4Cl + 20\,NH_3 \xrightarrow[\text{charcoal}]{O_2 \text{ (air)}} 4[Co(NH_3)_6]Cl_3 + 2\,H_2O$$

The technique of air bubbling is applicable to the oxidation of a cobalt(II) salt in a basic aqueous solution containing an amine (or ammonia) and a salt of the amine (or ammonia):

$$4\,CoCl_2 + 8\,C_2H_4(NH_2)_2 + 4\,C_2H_4(NH_2)_2 \cdot HCl \xrightarrow{O_2 \text{ (air)}}$$
$$4\,[Co(en)_3]\,Cl_3 + 2\,H_2O$$

$$4\,Co(NO_3)_2 + 16\,NH_3 + 4\,(NH_4)_2CO_3 \xrightarrow{O_2 \text{ (air)}}$$
$$4\,[Co(CO_3)(NH_3)_5]\,NO_3 + 4\,NH_4NO_3 + 2\,H_2O$$

$$4\,CoCl_2 + 8\,C_2H_4(NH_2)_2 + 8\,HCl \xrightarrow{O_2 \text{ (air)}}$$
$$4\,trans\text{-}[CoCl_2(en)_2]\,Cl \cdot HCl + 2\,H_2O$$

2) Tris(oxalato)cobaltate(III) complex is prepared from the components using lead dioxide as the oxidizing agent [19,20].

$$CoCO_3 + H_2C_2O_4 \rightarrow CoC_2O_4 + H_2O + CO_2$$

$$2\,CoC_2O_4 + 4\,K_2C_2O_4 + PbO_2 + 4\,HC_2H_3O_2 \rightarrow$$

$$2\,K_3[Co(ox)_3] + 2\,KC_2H_3O_2 + Pb(C_2H_3O_2)_2 + 2\,H_2O$$

This method of PbO_2 oxidation has been revived recently for preparing mixed ligand complexes with such chromophores as $[Co(N)_2(O)_4]^-$ [21] and $[Co(N)_4(O)_2]^+$ [22]. 3) Anionopentaammine complexes, $[Co(a)(NH_3)_5]^{2+}$, can be prepared from a penta-amminecarbonato complex [23].

$$[Co(CO_3)(NH_3)_5]^+ \begin{cases} \xrightarrow{\text{NH}_4\text{HF}_2} [CoF(NH_3)_5]^{2+} \\ \xrightarrow{\text{HI}} [CoI(NH_3)_5]^{2+} \\ \xrightarrow[\text{HNO}_3]{\text{NaNO}_2} [Co(NO_2)(NH_3)_5]^{2+} \\ \xrightarrow{\text{HNO}_3} [Co(NO_3)(NH_3)_5]^{2+} \\ \xrightarrow{\text{CH}_3\text{COOH}} [Co(CH_3COO)(NH_3)_5]^{2+} \end{cases}$$

Isotopically labeled oxygen has proved that $[Co(CO_3)(NH_3)_5]^+$ acid-hydrolyzes to $[Co(NH_3)_5(H_2O)]^{3+}$ with O—C bond rupture in Co^{III}—O—CO_2 [24]. Fujita et al. [25] employed the $[Co(NH_3)_5(H_2O)]^{3+}$ complex in preparing L-amino acid complexes, $[Co(NH_3)_5(L-Ham)]\,X_3$, in which L-amino acid acts as a unidentate ligand.
4) cis-bis(ethylenediamine)dinitrocobalt(III) complex is usually prepared by the classical method of Werner [26] from potassium hexanitrocobaltate(III)

$$K_3[Co(NO_2)_6] + 2\,C_2H_4(NH_2)_2 \rightarrow$$
$$cis\text{-}[Co(NO_2)_2(en)_2]\,NO_2 + 3\,KNO_2$$

The hexanitrocobaltate(III) is now often used as a starting material, where the ligating NO_2^- ions are replaced partially or totally by an aimed ligand. Examples are $[Co(NO_2)_3(dien)]$ [27], trans-$[Co(NO_2)_2(2,4\text{-ptn})_2]^+$ [28], $[Co(RR\text{-}2,4\text{-ptnta})]^-$ [29], and trans-$[Co(NO_2)_2(acac)_2]^-$ [30], (dien = diethylenetriamine, 2,4-ptn = 2,4-pentane-diamine, 2,4-ptnta = 2,4-pentanediaminetetraacetate and acac = acetylacetonate).

Thus, the methods mentioned above are successfully applicable for new complexes with slight modifications of the original procedures.

1.1.3 Feature of Modern Coordination Chemistry

In the past 30 years, the knowledge of transition metal complexes has grown tremendously. Some reasons for this growth are: Instruments in electronic absorption spectroscopy, vibrational spectroscopy, X-ray spectroscopy, optical rotatory dispersion, circular dichroism, nuclear magnetic resonance, etc. are commercially available. A theoretical concept, the ligand-field theory has given an impetus to the chemistry of transition metal complexes. With this theory, the problems of optical properties, magnetic properties, and kinetic and thermodynamic stability can be interpreted effectively. The progress in design and synthesis of multidentate ligands has helped to develop the chelate chemistry. And the understanding of the structures and properties of biologically important substances may be effectively approached through knowledge of chelate complexes synthesized as models.

In addition, the preparative work advanced steadily, although the preparations are mainly described in the experimental parts of published papers.

In the following sections some fundamental methods of general syntheses from recent years will be described.

1.2 Complexes with Flexible Quadridentate Ligands

1.2.1 Stereochemical Significance of the $[Coa_2(trien)]^{n+}$-Type Complexes

A molecule of triethylenetetramine $NH_2CH_2CH_2NHCH_2CH_2NHCH_2CH_2NH_2$ (3,6-diaza-1,8-octanediamine, trien) coordinates to a Co(III) ion in three geometrical configurations, cis-α (s-cis), cis-β (u-cis) and trans (Fig. 1.1). Moreover, when two asymmetric secondary nitrogen atoms are present the isomeric possibilities in the $[Coa_2(trien)]^{n+}$ system (a = an aniono or a neutral ligand) increase. Stereo models demonstrate that in the cis-α configuration two isomers, Δ(SS) and Λ(RR), in the cis-β configuration four isomers, Λ(SS), Λ(SR), Δ(RR) and Δ(RS), and in the trans configuration three isomers, (SS), (RS) and (RR), possibly exist (Fig. 1.2). To elucidate these stereochemical views, cobalt(III) complexes with trien or its substituted derivatives have been extensively studied since the late sixties. Developments in this field of research up to 1970 have been reviewed by Brubaker and Schaefer [31].

In 1948, Basolo [32] prepared $[CoCl_2(trien)]^+$ and $[Co(NO_2)_2(trien)]^+$. Inspite of several different preparative methods only one kind of the dichloro complex was obtained. Because of its blue-violet color he assigned to it a cis configuration. Likewise he obtained only one dinitro compound, although several methods were used for the preparation. Later, Das Sarma and Bailar [33] observed that a purple solution of $[CoCl_2(trien)]$ Cl in methanol slowly changes to light gray-violet when left to stand, and they estimated from spectral data that the species in the gray-violet solution are ca. 40% trans-isomer and 60% cis-isomer. They also established that the

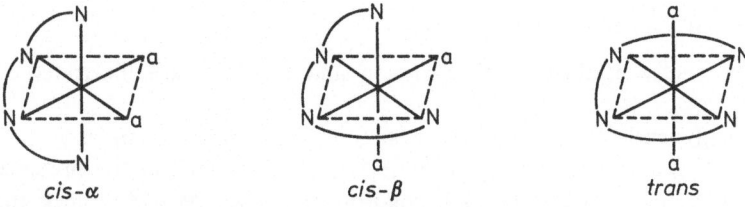

cis-α   cis-β   trans

Fig. 1.1. Geometrical forms of the $[Coa_2(trien)]^{n+}$ ion

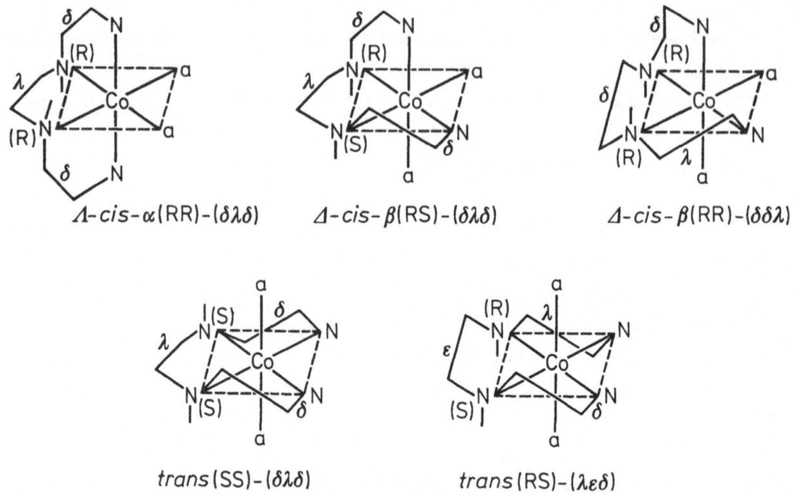

Fig. 1.2. Possible isomers of the $[Coa_2(trien)]^{n+}$ ion

blue-violet $[CoCl_2(trien)]$ Cl has the *cis* configuration by resolving the cation into optically active forms.

With the idea that the large sized bromide ion helps in forcing the planar configuration of trien, Selbin and Bailar [34] succeeded in isolating both green and the violet forms of the $[CoBr_2(trien)]$ Br, but Buckingham and Jones [35] concluded from IR spectra of cobalt(III) triethylenetetramine complexes that the green *trans* product is mainly *cis*-α contaminated by *cis*-β form. Gillard and Wilkinson [36] produced *trans*-$[CoCl_2(trien)]$ Cl simply by heating either moist *cis*-$[CoCl_2(trien)]$ Cl or *cis*-$[CoCl(H_2O)(trien)]$ Cl_2 to 180 °C, the aquachloro compound being obtained from a solution of the *cis*-dichloro compound acidified with HCl. However, Buckingham et al. [37] reported that this experiment gives rise chiefly to the reduction to cobalt(II).

In 1967, Sargeson and Searle [38] obtained all three geometrical isomers of $[CoCl_2$-(trien)] Cl and resolved the two *cis* forms into optically active antipodes. The *cis*-α and *cis*-β complexes containing 2 NO_2^-, 2 H_2O, or CO_3^{2-} were also systematically prepared and their optically active forms were gained. From these preparations stereochemical studies of complexes with flexible linear quadridentate ligands, in which the alkyl chains connecting the donor atoms were varied in length or in the positions of contained methyl groups.

1.2.2 Complexes of *cis*-α Series

The preparative routes established by Sargeson and Searle [38] are illustrated in Scheme 1.

cis-α-$[Co(NO_2)_2(trien)]$ Cl · H_2O. At the beginning, this dinitro complex is prepared by air-oxidation of aqueous $CoCl_2 · 6 H_2O$ and trien · HCl in the presence of excess $NaNO_2$; the mixed solution is vigorously aerated at 0 °C. Since the precipitated product contains some *cis*-β isomer, it is fractionally recrystallized

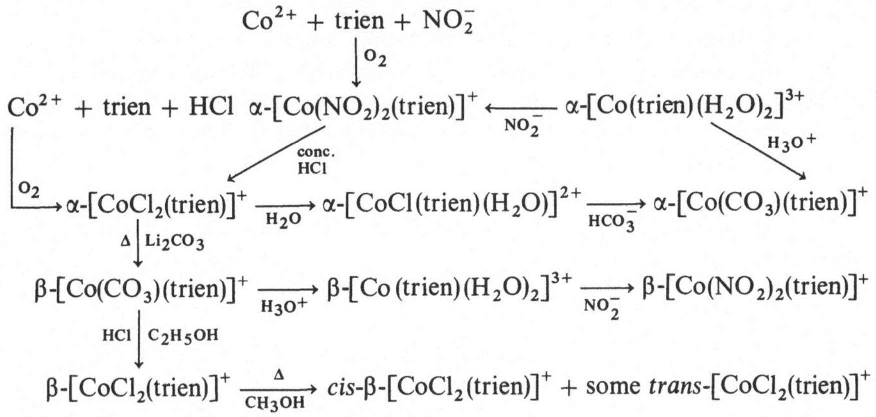

$$Co^{2+} + trien + NO_2^-$$

$$\downarrow O_2$$

$Co^{2+} + trien + HCl \quad \alpha\text{-}[Co(NO_2)_2(trien)]^+ \xleftarrow[NO_2^-]{} \alpha\text{-}[Co(trien)(H_2O)_2]^{3+}$

$$\downarrow O_2 \qquad\qquad\qquad \swarrow \underset{HCl}{\overset{conc.}{}} \qquad\qquad\qquad \searrow H_3O^+$$

$\alpha\text{-}[CoCl_2(trien)]^+ \xrightarrow[H_2O]{} \alpha\text{-}[CoCl(trien)(H_2O)]^{2+} \xrightarrow[HCO_3^-]{} \alpha\text{-}[Co(CO_3)(trien)]^+$

$$\Delta \downarrow Li_2CO_3$$

$\beta\text{-}[Co(CO_3)(trien)]^+ \xrightarrow[H_3O^+]{} \beta\text{-}[Co(trien)(H_2O)_2]^{3+} \xrightarrow[NO_2^-]{} \beta\text{-}[Co(NO_2)_2(trien)]^+$

$$HCl \downarrow C_2H_5OH$$

$\beta\text{-}[CoCl_2(trien)]^+ \xrightarrow[CH_3OH]{\Delta} cis\text{-}\beta\text{-}[CoCl_2(trien)]^+ + \text{some } trans\text{-}[CoCl_2(trien)]^+$

Scheme 1. Preparative Routes for trien-Co(III) Complexes

from hot water (90 °C), the least soluble material being *cis-α* isomer. The geometrical purity of the products is examined by paper chromatography in 1-butanol—pyridine—water—acetic acid (4:3:2:1) or 1-butanol—water—acetic acid (7:2:1). The absorption spectra of both isomers are of little use in distinguishing the isomers, but the IR spectra are usable; as to the NH absorptions $\sim 3300 \text{ cm}^{-1}$, the less symmetrical β isomer gives a more complex spectrum.

The corresponding *cis-α*-dinitro bromide, nitrate and acetate are prepared in a similar manner by using the respective cobalt(II) salts and acids. The analogous preparation of perchlorate complex gives a mixture of *cis-α* and *cis-β* isomers, with reversed order of solubilities.

cis-α-[CoCl$_2$(trien)] Cl. This purple compound is derived from *cis-α*-[Co(NO$_2$)$_2$-(trien)] Cl · H$_2$O by treatment with conc. HCl; when the dinitro chloride is dissolved in water and the excess HCl is evaporated to dryness, a crude product containing a small amount of *cis-β* isomer is obtained; this is recrystallized from minimum volume of boiling 3 mol/dm³ HCl to yield pure *cis-α* chloride.

According to Basolo, aerating a mixture of trien and CoCl$_2$, adding 10 mol/dm³ HCl and evaporating the solution, Sargeson and Searle obtained a crude product which was blue rather than violet and finally obtained pure *cis-α* dichloro chloride after repeated recrystallizations from boiling 3 mol/dm³ HCl.

Cis-α-[Co(CO$_3$)(trien)] ClO$_4$ · H$_2$O. Pure *cis-α* dichloro complex is added to a large volume of 0.013 mol/dm³ HClO$_4$ and the solution is stirred until the dichloro chloride is dissolved and aquated to the aquachloro stage (more than one day). Then, excess NaHCO$_3$ and NaClO$_4$ are successively added. When the mixture is evaporated at room temperature, the perchlorate monohydrate crystallizes out.

1.2.3 Complexes of *cis-β* Sereies

Cis-β-[Co(CO$_3$)(trien)] Cl · 1.5 H$_2$O. Treatment of *cis-α*-[CoCl$_2$(trien)] Cl with CO$_3^{2-}$ in basic solution produces the *cis-β* carbonato complex; a mixture of crude *cis-α* dichloro chloride and Li$_2$CO$_3$ in water is heated with stirring, whereupon the violet color changes to red and the solution becomes alkaline (after *ca.* 1 hr.). The excess Li_2CO_3 is filtered off when hot; after cooling, some $CaCl_2$ is added to eliminate CO_3^{2-}; the solution is then filtered. The compound is obtained by adding ethanol to the filtrate and then leaving the solution for some time. The compound is recrystallized from a warm solution by adding ethanol. This carbonato complex is used for the preparation of other complexes of *cis-β* series through simple substitution reactions.

In order to distinguish α- and β-carbonato complexes, absorption spectral data are useful (Fig. 1.3).

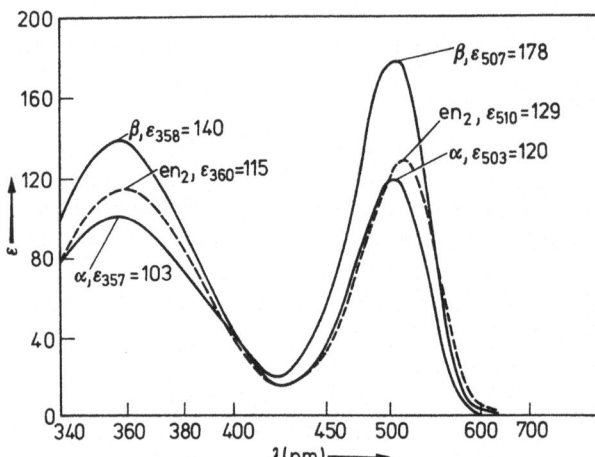

Fig. 1.3. Absorption spectra for α- and β-[Co(CO₃)(trien)]⁺ and [Co(CO₃)(en)₂]⁺ in aqueous solutions (from Ref. [38])

Cis-β-[Co(NO₂)₂(trien)] NO₃. This is derived from the β-carbonato complex through two reaction steps; a slight excess of 2 mol/dm³ HNO_3 is added to the carbonato chloride to produce diaqua complex species. After *ca.* 30 min. $NaNO_2$ is added. Yellow-orange crystals slowly separate and are recrystallized from hot water.

Cis-β-[CoCl₂(trien)] Cl · 1.5 H_2O. This red-violet compound is prepared from the β-carbonato chloride; it is suspended in ethanol saturated with dry hydrogen chloride and stirred overnight; then the compound precipitates. A saturated solution of this crude compound in dil. acetic acid (pH 3) is prepared quickly at room temperature; then the solution is quickly cooled in ice and conc. HCl is added for recrystallization.

1.2.4 *Trans*-Dichloro Complex

The major difference between the trien complexes and the corresponding bis(ethylenediamine) complexes is the instability of the *trans*-isomer in the former. The *trans*-dichloro complex can be prepared only from *cis-β* dichloro complex.

Trans-[CoCl₂(trien)] ClO₄. Crude *cis-β* dichloro chloride is refluxed in a large volume of methanol (*ca.* 1 hr.). The residue is filtered off and excess dry $LiClO_4$ is added to the green solution. On cooling in ice and scratching the vessel, the perchlorate of the *trans*-dichloro complex is obtained in green crystals. These are filtered off after *ca.* 10 min. to avoid contamination with the *cis*-isomers. The residue is extracted again and another crop of the *trans*-isomer is obtained.

Trans-[CoCl$_2$(trien)] Cl [37)]. Finely powdered *cis*-β dichloro chloride is suspended in methanol and is made slightly acidic with dry HCl gas. The suspension is heated under reflux and stirring. After 20–30 min. intervals the undissolved β-dichloro chloride is filtered and extracted with fresh methanol. The brown-green filtrate is concentrated under reduced pressure until the desired compound begins to crystallize.

The visible absorption spectra and the IR spectra in the NH asymmetric deformation region for the α-, β- and *trans*-dichloro isomeric complexes have been reported [35)].

The following concepts will be useful for the syntheses of other complexes, from the above-mentioned systematic synthesis:

i) The coordinated NO$_2^-$ in a complex tends to be substituted by Cl$^-$ by treatment with conc. HCl, and the produced chloro complex is converted to aqua complex species by acidifying the solution. From this fact, a reaction route from a nitro complex, via chloro and then aqua, to a desired complex seems to be a useful pathway in synthesis.

ii) The conversion, *cis*-dichloro→aquachloro→diaqua, in dilute acid solution takes place with the retention of configuration. The same applies to the acid hydrolysis of a carbonato complex. Therefore, these reactions serve to produce complexes with the same geometrical structure as the starting material.

iii) A carbonato complex when treated with ethanolic (or methanolic) HCl results in the corresponding *cis*-dichloro complex. With conc. HCl it produces some *trans*-dichloro complex.

1.2.5 Optical Resolutions

Sargeson and Searle prepared optically active isomeric complexes either by diastereo-isomeric salt formation or by derivation from an optically active complex; the *cis*-α-dinitro complex and the *cis*-α-dichloro complex are resolved with Na(−) [Co(ox)$_2$(en)] H$_2$O and Na(+) [Co(ox)$_2$(en)] · H$_2$O, respectively. Resolution of the *cis*-β-dinitro complex is achieved with Na$_2$[Sb$_2$(*d*-tart)$_2$]. *Cis*-β-dichloro complex is resolved with Na(+) [Co(ox)$_2$(en)] · H$_2$O at a low temperature, since the complex aquates rapidly (t$_{1/2}$ in acid solution is 8 min at 25 °C).

(+)α-[Co(CO$_3$)(trien)] ClO$_4$ · H$_2$O is derived from (+)*cis*-α-[CoCl$_2$(trien)] ClO$_4$. This dichloro complex is aquated in 0.04 mol/dm^3 HClO$_4$; by adding NaHCO$_3$ the carbonato complex begins to crystallize. Aquation of the (+)α-[Co(CO$_3$)(trien)]$^+$ with aqueous HClO$_4$ at 0 °C results in the formation of (+)$_{546}$*cis*-α-[Co(trien)(H$_2$O)$_2$]$^{3+}$.

The compound, β-[Co(CO$_3$)(trien)] Br, is resolved with (+)Ag-[Co(mal)$_2$(en)] · 2 H$_2$O. The aquation of (−)β-[Co(CO$_3$)(trien)]$^+$ with aqueous HClO$_4$ gives (−)$_{546}$*cis*-β-[Co(trien)(H$_2$O)$_2$]$^{3+}$.

The preparation of the active *trans*-dichloro complex was achieved [37)] according to the following reaction:

$$(+)β\text{-SS-}[CoCl_2(trien)]ClO_4 \xrightarrow[\text{reflux}]{\text{CH}_3\text{OH (trace HCl)}} trans\text{-SS-}[CoCl_2(trien)]ClO_4$$

$$[M]_D + 880° \qquad\qquad\qquad [M]_D + 3000°$$

Namely, (+)β-dichloro perchlorate is dissolved in anhydrous methanol acidified with dry HCl gas. The solution is heated under reflux for some time, cooled to room temperature, and filtered. Excess LiClO₄ is added to the filtrate, and after scratching the vessel at 0 °C, light green crystals separate out. The extraction procedure is repeated with the remaining (+)β-dichloro perchlorate. The crude product thus obtained is purified by converting it to chloride with tetraphenylarsonium chloride and again to original perchlorate with NaClO₄. A similar experiment using (−)β-RR-[CoCl₂(trien)] ClO₄ gives *trans*-RR-[CoCl₂(trien)] ClO₄.

1.2.6 Complexes with other Tetramine Ligands

In contrast to the complexes with trien, complexes with 3,7-diaza-1,9-nonanediamine, NH₂CH₂CH₂NHCH₂CH₂CH₂NHCH₂CH₂NH₂ (2,3,2-tet) markedly prefer the *trans* geometry. Aerial oxidation synthesis produced only *trans* complex [39,40]; an aqueous solution containing CoCl₂ · 6 H₂O and the tetramine is oxidized by air which has been passed through an aqueous NaOH solution and then dried. Subsequently conc. HCl is added. On evaporating the resulting green solution, crystals of *trans*(RS)-[CoCl₂(2,3,2-tet)] Cl are obtained [39].

The ligand 4,7-diaza-1,10-decanediamine, NH₂CH₂CH₂CH₂NHCH₂CH₂NHCH₂-CH₂CH₂NH₂ (3,2,3-tet) coordinates to cobalt(III) preferably with *trans* geometry [41]; the air oxidation synthesis gives *trans* (RR, SS racemate)-[CoCl₂(3,2,3-tet)] ClO₄.

The other *trans*(SS or RR)-complexes were prepared by Brubaker and Schaefer [42]; the complexes, β-[Co(ox)(2,3,2-tet)]⁺ and α-[Co(ox)(3,2,3-tet)]⁺, are prepared by heating an aqueous solution of each *trans*-dichloro complex [39,41] with an excess of potassium oxalate. The two complex ions thus prepared are fractioanally crystallized as both the iodide and the chloride salts. There is only one geometrical isomer produced for each of the complexes. These oxalato complexes are resolved with lithium hydrogen *d*-tartrate. (−)-*trans*(RR)-[CoCl₂(2,3,2-tet)] ClO₄ is prepared by refluxing (−)-β-[Co(ox)(2,3,2-tet)] ClO₄ in methanol saturated with dry HCl gas. Similarly, (−)-*trans*(RR)-[CoCl₂(3,2,3-tet)] ClO₄ is prepared from (+)₅₄₆-α-[Co(ox)-(3,2,3-tet)] ClO₄.

The stereochemistry of dianionocobalt(III) complexes with methyl-substituted trien [43−47] or with 2,3,2-tet derivatives [48,49] have been extensively elucidated by Yoshikawa and co-workers.

1.2.7 Complexes with Dithiadiamine Ligands

The preparation of dianiono complexes with 1,8-diamino-3,6-dithiaoctane NH₂CH₂-CH₂SCH₂CH₂SCH₂CH₂NH₂ (eee) was reported by Worrell and Busch [50]; it is very similar to that for the trien complexes. At first [Co(NO₂)₂(eee)] Cl is prepared by aerating a mixture of CoCl₂ · 6 H₂O, eee · HCl and NaNO₂ in methanol. The dinitro complex is converted to [CoCl₂(eee)] Cl by warming it in conc. HCl. However, air oxidation of a mixture of CoCl₂ · 6 H₂O, eee · HCl in water, methanol or ethanol-water failed to produce crystalline dichloro chloride, whereas air oxidation of aqueous or alcoholic solution of CoBr₂ · 6 H₂O and eee · HBr resulted in [CoBr₂(eee)] Br. Either Li₂CO₃ or NaHCO₃ reacts with the dichloro (or dibromo) complex to produce [Co(CO₃)(eee)]⁺, which is most easily crystallized as perchlorate. Although three geometric isomers are possible, all the complexes were characterized to be *cis-α*. The optical resolutions of the *cis-α*-dinitro complex and the *cis-α*-dichloro complex are achieved with Na(—)-[Co(ox)$_2$(en)] · H$_2$O and K$_2$[Sb$_2$(*d*-tart)$_2$] respectively [51]. The optically active α-carbonato complex is derived from the active dichloro complex *via* the active aquachloro complex.

Bosnich et al. [52] used related ligands such as 1,9-diamino-3,7-dithianonane NH$_2$CH$_2$CH$_2$SCH$_2$CH$_2$CH$_2$SCH$_2$CH$_2$NH$_2$ (ete) and 1,10-diamino-4,7-dithiadecane, NH$_2$CH$_2$CH$_2$CH$_2$SCH$_2$CH$_2$SCH$_2$CH$_2$CH$_2$NH$_2$ (tet) to prepare isomeric dichloro complexes. Air oxidation is done with an ice-cold solution of Co(OAc)$_2$ · 4 H$_2$O, ete · HCl (or tet · HCl) and NaNO$_2$ in methanol-water to obtain deposits of the dinitro complex, which is then treated with hydrochloric acid. From the ete ligand, *trans* and *cis-β* dichloro isomers are obtained, while from the tet ligand, *cis-α* and *cis-β* isomers are obtained. By the way, only the *cis-α* isomer is formed after aerating a methanol solution, followed by the addition of conc. HCl. A series of *cis-α*-[Coa$_2$-(eee)]$^{n+}$ complexes (a = Br$^-$, N$_3^-$, NCS$^-$, and NO$_2^-$; a$_2$ = C$_2$O$_4^{2-}$ and bpy) can be derived from *cis-α*-[CoCl$_2$(eee)] ClO$_4$.

1.3 Complexes with Macrocyclic Quadridentate Ligand

Since the discovery that vitamin B$_{12}$ is a macrocyclic complex of Co(III), there has been considerable interest in the preparation and characterization of analogous synthetic complexes. For example, Schrauzer and co-workers [53–55] prepared bis(dimethylglyoximato)cobalt complexes, which chemically closely resemble vitamin B$_{12}$ derivatives. The preparation of compounds such as aniono(pyridine)cobaloxime(III), [Co(a)(DH)$_2$(py)] (a = Cl$^-$, CN$^-$, N$_3^-$, NCS$^-$, Br$^-$, I$^-$; DH = monoanion of hydrogendimethylglyoximate), diaquacobaloxime(II), [Co(H$_2$O)$_2$(DH)$_2$], methyl-(pyridine)cobaloxime, CH$_3$Co(HD)$_2$py, methylaquacobaloxime, CH$_3$Co(DH)$_2$H$_2$O, and phenyl(pyridine)cobaloxime, C$_6$H$_5$Co(DH)$_2$py, has been published in Inorganic Syntheses [56], where the term "cobaloxime" is used for the bis(dimethylglyoximato)-cobalt moiety (see Fig. 1.4). These studies have revealed that the cobalt atom in the square planar ligand field of the four nitrogen atoms of dimethylglyoxime has a pronounced tendency to form stable organocobalt derivatives.

Sadasivan et al. [57] studied cobalt(III) complexes containing 5,7,7,12,14,14-hexamethyl-1,4,8,11-tetraazacyclotetradeca-4,11-dien, A, the Cu(II) and Ni(II) complexes of which had been extensively investigated by Curtis and co-workers [58–61].

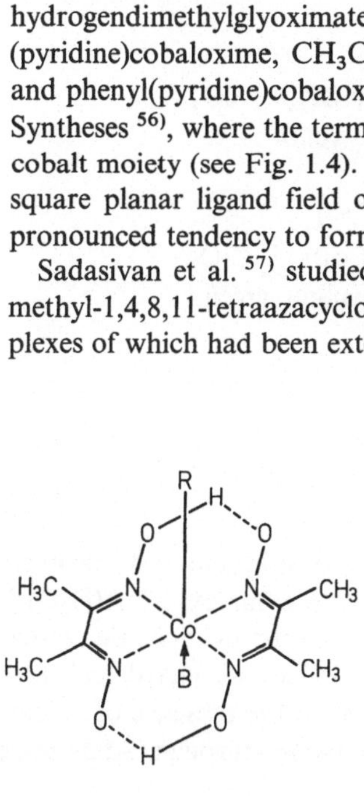

Fig. 1.4. Bis(dimethylglyoximato)cobalt complexes

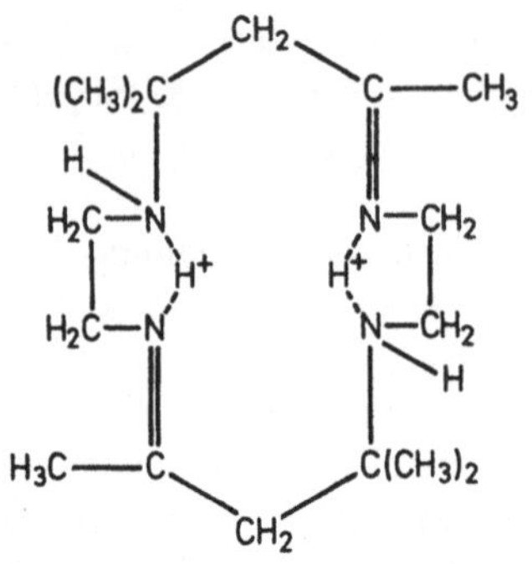

A

Scheme 2 summarizes the preparative routes to the $[Coa_2A]^{n+}$-type complexes. The ligand, as the hydrogen perchlorate salt $A \cdot 2\,HClO_4$, is prepared by reaction of $[Fe(en)_3]\,(ClO_4)_2$ with acetone [62]. At first, the carbonato complex, $[Co(CO_3)\,A]^+$ is prepared by air-oxidizing a mixture of $CoCO_3$ and a slight excess of $A \cdot 2\,HClO_4$ in aqueous methanol. The same compound is prepared by reacting $A \cdot 2\,HClO_4$ with $Na_3[Co(CO_3)_3] \cdot 3\,H_2O$ in water. Treatment of $[Co(CO_3)\,A]^+$ with conc. HCl and evaporation of the solution result in crystallization of the dichloro complex, $[CoCl_2A]\,ClO_4$. An excess of the dichloro compound is added to water (50 °C), the mixture is filtered, and the filtrate is mixed with some conc. $HClO_4$. After one week the aquachloro complex, $[CoClA(H_2O)]\,(ClO_4)_2$, crystallizes out.

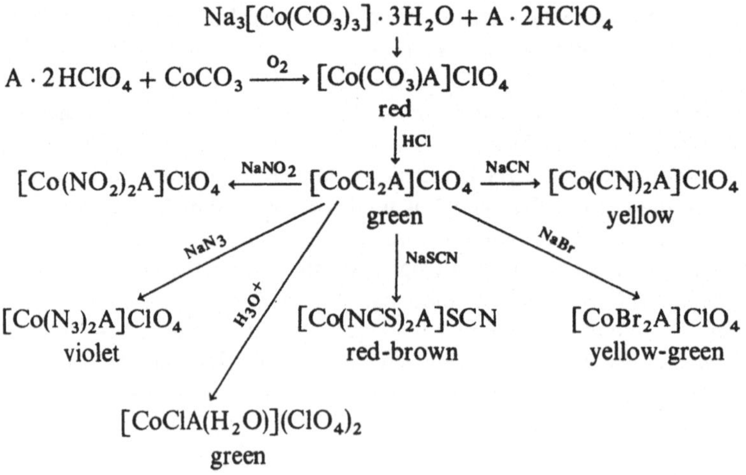

Scheme 2. Preparative Routes to $[Coa_2A]^{n+}$ Complexes

The other complexes such as dicyano, diazido, dinitro, dibromo and diisothiocyanato complexes are prepared analogously. The diaqua complex, $[CoA(H_2O)_2]$-$(ClO_4)_3$, is synthesized from the free amine prepared by Curtis method. Equimolar ethanolic solutions of the free amine and $Co(ClO_4)_2 \cdot 6\,H_2O$ are mixed and after filtering out the precipitate, the mixed solution is boiled in the presence of "Charcoal Activated Norit." While boiling, air or oxygen is passed through. After some time, the filtered mixture is acidified with $HClO_4$ to give a 30% solution. Then the diaqua complex slowly forms fine crystals.

The IR spectra, chemical properties and visible spectra showed that the unidentate ligands are *trans* to each other, indicating square planar coordination of the Schiff base macrocycle.

The ligands, 1,4,8,11-tetrathiacyclotetradecane (TTP), *I* and 13,14-benzo-1,4,8,11-tetrathiacyclopentadecane(TTX), *II*, are 14-membered and 15-membered quadridentate ligands, respectively, having four thioethers as donors. Macrocyclic ligands that are quite flexibly coordinate to a metal ion in a folded manner when the central metal ion is large. The critical ring size differs for different sets of donor atoms and metal ions. In this respect, Travis and Busch [63] prepared a series of cobalt(III) complexes with TTP or TTX. They found that the 14-membered ring, TTP, coordinates to Co(III) in a folded manner to give *cis*-$[Coa_2(TTP)]^+$ when a = Cl^-, Br^-, NCS^-, NO_2^- and $^1/_2 C_2O_4^{2-}$ and is forced to coordinate to Co(III) with a planar array of the cyclic donor atoms to give *trans*-$[CoI_2(TTP)]^+$ when a = I^-. The 15-membered ring, TTX, coordinates to Co(III) to give *trans*-$[Coa_2(TTX)]^+$ when a = Cl^- and Br^-.

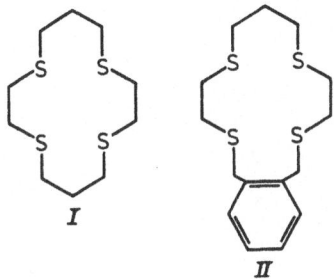

I

II

Scheme 3 illustrates preparative routes of the TTP complexes. First of all, hexaacetonitrilecobalt(II) tetrafluoroborate, $[Co(CH_3CN)_6] (BF_4)_2$, is prepared according to Hathaway et al. [64]; an excess metallic cobalt reacts with nitrosyl tetrafluoroborate in acetonitrile; the evolving nitric oxide is evacuated. The filtrated reaction mixture is concentrated and then anhydrous ether is added to isolate orange crystals. This complex is unstable in water or ethanol so that it dissociates to the free ligand and metal ion. In nitromethane, however, it is quite stable. Thus, the reaction of the complex with an excess of TTP in nitromethane at room temperature gives a microcrystalline product, $[Co(TTP)](BF_4)_2$ (red-brown), after adding ether.

The above Co(II) derivative has a noncoordinating anion (BF_4^-) and is quite stable in nitromethane toward oxidation until coordinating anions are introduced into the system. Thus, when lithium chloride is added to a nitromethane solution of $[Co(TTP)] (BF_4)_2$ and exposed to air, the resulting solution gives *cis*-$[CoCl_2(TTP)]$-BF_4 after adding ether. The bromide derivative, *cis*-$[CoBr_2(TTP)] BF_4$, is prepared in a similar manner.

The same dichloro complex is prepared by air-oxidation of the components in a one-step procedure; either cobalt(II) perchlorate hexahydrate and excess lithium chloride or anhydrous cobalt(II) chloride and excess lithium perchlorate are stirred

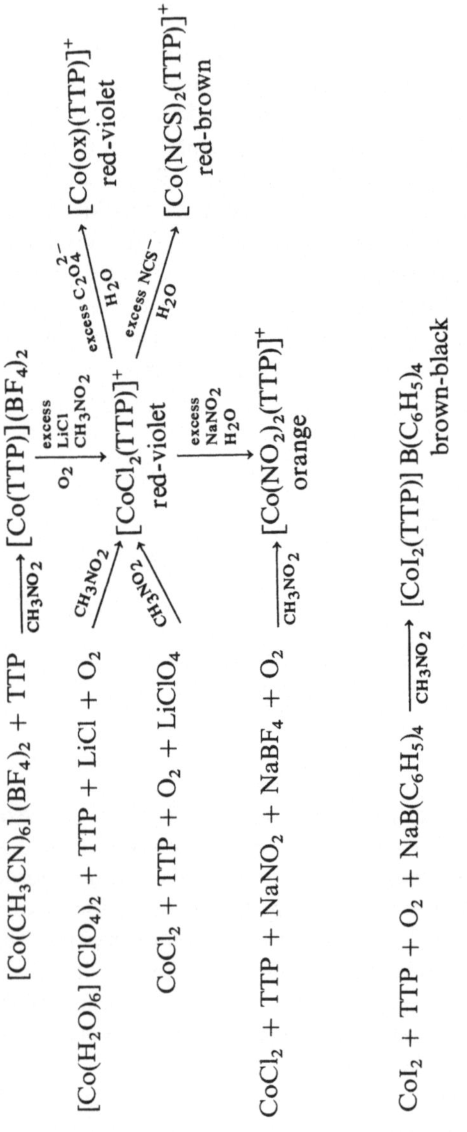

$$[Co(CH_3CN)_6](BF_4)_2 + TTP \xrightarrow{CH_3NO_2} [Co(TTP)](BF_4)_2$$

$$[Co(H_2O)_6](ClO_4)_2 + TTP + LiCl + O_2 \xrightarrow{CH_3NO_2}$$

$$CoCl_2 + TTP + O_2 + LiClO_4 \xrightarrow{CH_3NO_2} [CoCl_2(TTP)]^+ \quad \text{red-violet}$$

$[Co(TTP)](BF_4)_2 \xrightarrow[\text{CH}_3\text{NO}_2]{\text{excess LiCl}} [CoCl_2(TTP)]^+$ red-violet

$\xrightarrow[\text{H}_2\text{O}]{\text{excess } C_2O_4^{2-}} [Co(ox)(TTP)]^+$ red-violet

$\xrightarrow[\text{H}_2\text{O}]{\text{excess } NCS^-} [Co(NCS)_2(TTP)]^+$ red-brown

$[CoCl_2(TTP)]^+ \xrightarrow[\text{H}_2\text{O}]{\text{excess NaNO}_2} [Co(NO_2)_2(TTP)]^+$ orange

$$CoCl_2 + TTP + NaNO_2 + NaBF_4 + O_2 \xrightarrow{CH_3NO_2} [Co(NO_2)_2(TTP)]^+ \quad \text{orange}$$

$$CoI_2 + TTP + O_2 + NaB(C_6H_5)_4 \xrightarrow{CH_3NO_2} [CoI_2(TTP)] B(C_6H_5)_4 \quad \text{brown-black}$$

Scheme 3. Preparative Routes of Co(III)-TTP Complexes

together with a slight excess of the ligand. When ether is added the complex crystallizes as perchlorate.

The other dianiono complexes such as cis-[Co(NCS)$_2$(TTP)]$^+$, [Co(ox) (TTP)]$^+$ and cis-[Co(NO$_2$)$_2$(TTP)]$^+$ are derived from the cis-dichloro complex by analogous procedures; a solution of cis-[CoCl$_2$(TTP)] ClO$_4$ and excess sodium thiocyanate in water is stirred and refluxed for some time. When sodium tetraphenylborate is added, the diisothiocyanato complex separates as its tetraphenylborate. Similarly, from a refluxed solution of cis-dichloro complex perchlorate and sodium oxalate, the oxalato complex is obtained as its perchlorate. A mixed solution of cis-[CoCl$_2$(TTP)] BF$_4$ with excess sodium nitrite is boiled for some time, then the dinitro compound separates as its tetrafluoroborate. This dinitro compound is obtained by aeration of a mixture of the components in nitromethane, too. The dianiono complexes are stable against dissociation in all common solvents, and are recrystallized from hot acetonitrile.

The diiodo complex, trans-[CoI$_2$(TTP)]$^+$, is prepared by air oxidation; anhydrous cobalt(II) iodide is stirred with the ligand in nitromethane, whereby a cobalt complex containing tetrahedral CoI$_4^{2-}$ species is formed, which is then isolated by adding ether. The complex is redissolved in nitromethane and exposed to air for a few days. Successive additions of sodium tetraphenylborate and ether precipitates the desired compound.

The TTX complexes, trans-[CoCl$_2$(TTX)] (ClO$_4$) and trans-[CoBr$_2$(TTX)] ClO$_4$, are prepared by allowing [Co(TTX)] (ClO$_4$)$_2$ to react with an appropriate lithium halide in nitromethane. Reaction of [Co(TTX)] (ClO$_4$)$_2$ with lithium iodide produces the cobalt(II) complex, [CoI$_2$(TTX)].

Through the mentioned preparations, three general methods have been established [63]. 1) Preparation and isolation of the Co(II)-TTP complex with a non-coordinating anion, reaction of this complex with a coordinating anion, and then air oxidation; 2) Reaction of a Co(II) salt, the ligand, coordinating and/or non-coordinating anion, and air oxidation all in a one-step procedure; 3) Exchange of aniono groups of a previously isolated Co(III) complex for those of a more strongly coordinating anion by similar procedures.

1.4 Complexes Containing α,ω-Alkanediamines

Stable chelate rings of five and six members containing metal ions are well known but rings of seven or more members are uncommon. Since Werner's time, metal complexes having large chelate rings were attempted to synthesize; those studies up to 1950 have been reviewed by O'Brien [65]. He stated: "It is quite evident that the proposed structures of complexes with chelate rings containing more than six atoms are not firmly established. Lack of X-ray and other conclusive data, the several possible linkages, and the possibility of polymerization, all tend to make the proposed structures highly speculative".

The preparation of cobalt(III) complexes containing α,ω-alkanediamine H$_2$N-(CH$_2$)$_n$NH$_2$ (N—N) was reported in 1973 by Ogino and Fujita [66], who communicated the syntheses of bis(ethylenediamine)-tetramethylenediamine (tmd) and -dodeca-methylenediamine (don) complexes. Two years later, they synthesized [Co(en)$_2$-

$(H_2N(CH_2)_nNH_2)]X_3$ (n = 4, 10, 12 and 14) and $]Co_2(en)_4(H_2N(CH_2)_nNH_2)_2]X_6$ (n = 4, 5, 6, 8, 10 and 12) [67]. These complexes were obtained from the reaction products of cis- or trans-[CoCl_2(en)_2] Cl with the respective N—N ligand in dimethyl sulfoxide (DMSO). They recommended the cis-dichloro isomer as a better starting material because it is more soluble in DMSO than the trans-dichloro isomer.

[Co(en)_2(tmd)]Br_3 and [Co_2(en)_4(tmd)_2]Br_6 · 6 H_2O: A large volume (550 cm³) of DMSO containing cis-[CoCl_2(en)_2]Cl (0.015 mol) and tmd (0.016 mol) is kept at 50 °C for 20 hr. After cooling to room temperature, the solution is neutralized with conc. HCl and diluted to a large volume (3 l) with water. This is separated on a SP-Sephadex C-25 column (5 × 40 cm) with 0.1 mol/dm³ KBr into pink (very small amount), red (small amount), yellow and yellowish orange bands descending in this order. The first two bands are eluted with 0.2 mol/dm³ KBr. The third yellow band is eluted with 0.4 mol/dm³ KBr, yielding crystals of [Co(en)_2(tmd)] Br_3. The last yellowish orange band is eluted with a warm 1 mol/dm³ KBr solution and gives crystals of the dimeric complex [Co_2(en)_4(tmd)_2] Br_6 · 6 H_2O.

In the [Co(en)_2(N—N)]³⁺ complexes, two nitrogen atoms of the N—N can occupy either cis or trans position as shown in Fig. 1.5a, if the methylenic chain is long enough. However, from the fact that the [Co(en)_2(N—N)]³⁺ complexes (N—N = tmd and don) were resolved into optical isomers by SP-Sephadex column chromatography the workers assigned cis structure to all the complexes. For the dimeric complexes, [Co_2(en)_4(N—N)_2]⁶⁺, structures shown in Fig. 1.5b were proposed as most probable on the basis of molecular models.

The success in the syntheses is due to the use of DMSO as the solvent, in which very dilute concentration of the reactants predominantly gives rise to the desired product. The yield of [Co(en)_2(tmd)]³⁺ amounts to 70%, when 0.015 mol of tmd is allowed to react with an equivalent amount of [CoCl_2(en)_2]Cl in 550 cm³ of DMSO. while in 50 cm³ DMSO, the yield of the monomeric complex decreases to a few percent and that of the dimeric complex increases. The mole ratio of the starting complex to N—N also affects the distribution of the products. The reaction of 0.01 mol of [CoCl_2(en)_2]⁺ with 0.02 mol of tmd in 100 cm³ of DMSO gives a small

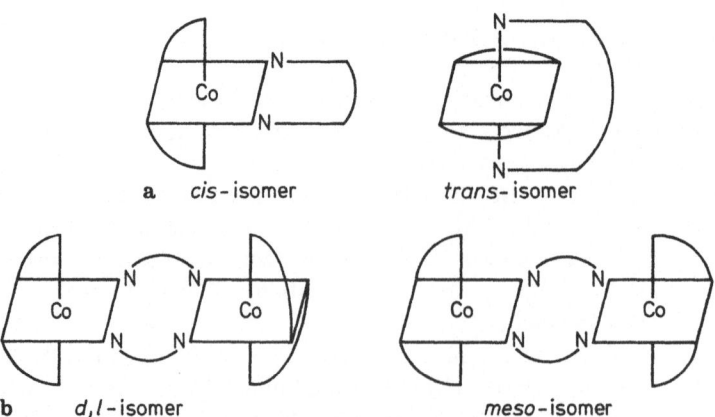

a   cis-isomer   trans-isomer

b   d,l-isomer   meso-isomer

Fig. 1.5a and b. Structures for the monomer and the dimer

amount of $[Co(en)_2(tmd)]^{3+}$ and large amounts of $[Co(en)_2(Htmd)_2]^{5+}$ and $[Co_2(en)_4$-$(tmd)_2]^{6+}$. From these experimental facts, the workers have proposed the following reaction mechanism;

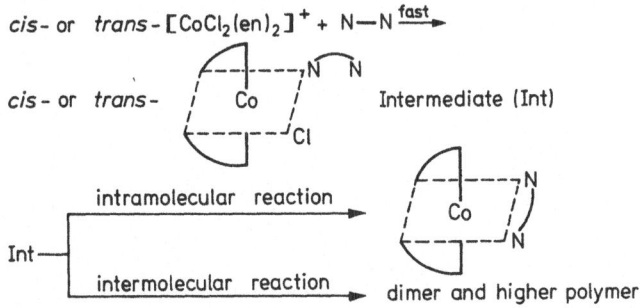

Furthermore, Ogino[68] examined reactions of $[Co(NH_3)_5(H_2O)](ClO_4)_3$ with $H_2N(CH_2)_nNH_2$ in DMSO by changing the number of methylene groups in N—N. The procedures for the isolation of complexes were similar to those for the above-mentioned complexes. The production of $[Co(NH_3)_4(N-N)]^{3+}$ containing an N—N chelate ring was observed for n = 2 ~ 4, 12 and 14, but not for n = 5, 7, 8 and 10, indicating that the medium-size chelate ring is unstable.

Fujita and Ogino[69] used DMSO for the synthesis of the tris(tetramethylene-diamine) complex, $[Co(tmd)_3]Br_3$. A DMSO solution containing $Co(NO_3)_2 \cdot 6 H_2O$, tmd $\cdot$ 2 HCl and free tmd in the mole ratio of 1:1:3 is aerated in the presence of activated charcoal, and the resulting solution is then diluted to a large volume with water and adjusted to pH *ca.* 3 with HCl. From this, the desired compound is obtained *via* SP-Sephadex C-25 column chromatography. The compound can be resolved with $Ag_2[Sb_2(d\text{-tart})_2]$. The absolute configuration of the $(-)_{589}$-enantiomer has been assigned to be Λ on the basis of CD sign.

1.5 Mixed Cyano Complexes

The cyanide ion occupies an extremely high position in the spectrochemical series an exhibits strong tendency to coordinate to cobalt(III). For the hexacyanocobalt-ate(III) complex, $K_3[Co(CN)_6]$, it is not necessary to aerate the solution of hexa-cyanocobaltate(II) formed from a cobalt(II) salt and excess of potassium cyanide, since the $[Co(CN)_6]^{4-}$ ion is oxidized by water with evolution of hydrogen. Namely, the stabilization of Co(III) against Co(II) is greatly favoured by the coordina-tion with cyanide ions.

Chan and Tobe[70] investigated the reaction between *trans*-$[CoCl_2(en)_2]^+$ and cyanide and found that the concentrated aqueous solution was disproportionated to $[Co(en)_3]^{3+}$ and some unidentifiable products, whereas in dilute solution hydrolysis of the cyanide ion occurred. This suggests that the design to replace a coordinated ligand such as Cl^- in a complex by CN^- is undesirable. However, mixed cyano complexes were obtained in the 1960's.

1.5.1 Methods Using Complexes with Sulfite or Thiosulfate Ion

Chan and Tobe [70] obtained *trans*-forms of monocyano complexes, *trans*-[Coa(CN)-(en)$_2$]$^{n+}$ (a = Cl$^-$, H$_2$O and OH$^-$), by the following reaction sequence:

$$trans\text{-}[CoCl_2(en)_2]^+ \xrightarrow{SO_3^{2-}} [CoCl(SO_3)(en)_2] \xrightarrow{CN^-}$$

$$[Co(CN)(SO_3)(en)_2] \xrightarrow{conc.\ HCl} trans\text{-}[CoCl(CN)(en)_2]^+ \rightarrow$$

$$\xrightarrow[Ag^+]{H_2O} trans\text{-}[Co(CN)(H_2O)(en)_2]^{2+}$$

$$\Big\uparrow H_3O^+$$

$$\xrightarrow{OH^-} trans\text{-}[Co(CN)(OH)(en)_2]^+$$

$$\xrightarrow{conc.\ HBr} trans\text{-}[CoBr(CN)(en)_2]^+ \ [71]$$

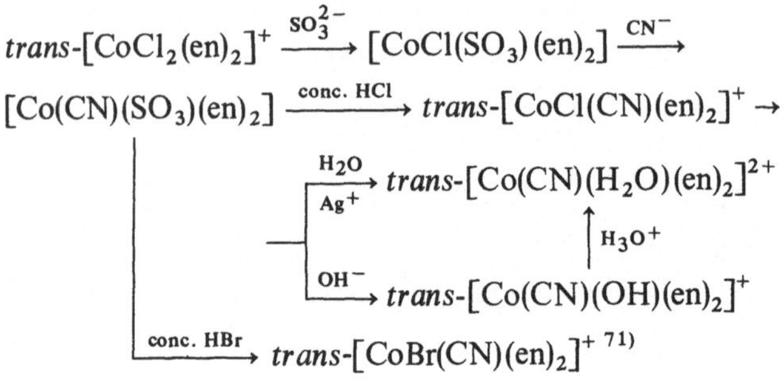

Ohkawa et al. [72] obtained *cis*-forms of the monocyano complexes, *cis*-[Coa(CN)-(en)$_2$]$^{n+}$ (a = Cl$^-$, Br$^-$, I$^-$, NO$_2^-$, H$_2$O and NH$_3$) by the following reaction sequence:

$$cis\text{-}[Co(SO_3)_2(NH_3)_4]^- \ [73,74]$$

$$\xrightarrow{CN^-}$$

$$\xrightarrow[heat]{CN^-}$$

dark yellow-[Co(CN)$_2$(SO$_3$)$_2$(NH$_3$)$_2$]$^-$

orange yellow-[Co(CN)$_2$(SO$_3$)$_2$(NH$_3$)$_2$]$^-$ $\xrightarrow[heat]{en}$

$$cis\text{-}[Co(CN)(SO_3)(en)_2] \ \begin{cases} \xrightarrow[heat]{conc.\ HCl} cis\text{-}[CoCl(CN)(en)_2]^+ \\ \xrightarrow[heat]{conc.\ HBr} cis\text{-}[CoBr(CN)(en)_2]^+ \end{cases}$$

$$\xrightarrow{Ag_2O,\ HI} cis\text{-}[Co(CN)I(en)_2]^+$$

$$\xrightarrow{NO_2^-} cis\text{-}[Co(CN)(NO_2)(en)_2]^+$$

$$\xrightarrow{liquid\ NH_3} cis\text{-}[Co(CN)(NH_3)(en)_2]^{2+}$$

$$\xrightarrow{Ag_2O,\ HBr} cis\text{-}[Co(CN)(en)_2(H_2O)]^{2+}$$

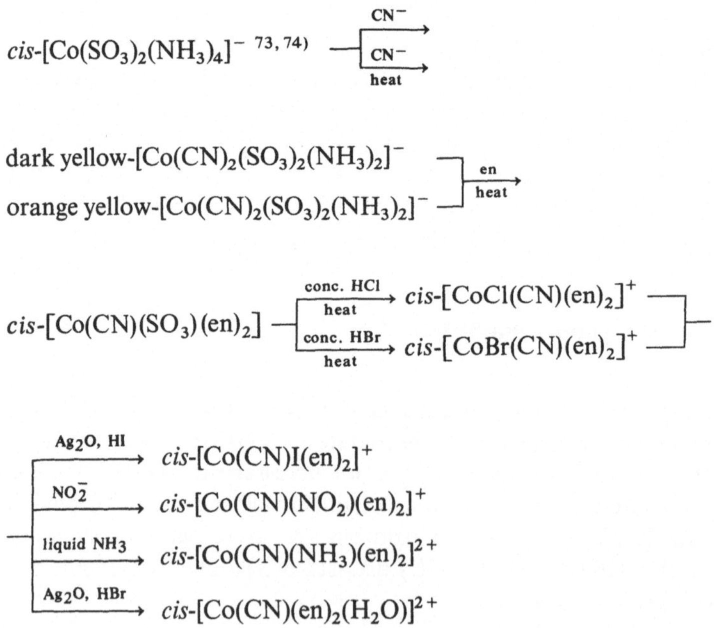

These *cis*-complexes were all resolved into their optically active isomers, using ammonium (+)-bromocamphor-π-sulfonate as resolving agent.

Siebert [75] obtained pentaamminecyano complexes [CoCN(NH$_3$)$_5$]X$_2$ (X = Cl$^-$, Br$^-$, NO$_3^-$, ClO$_4^-$ or $^1/_2$ SO$_4^{2-}$) by the following reactions;

i) Co^{2+}, NH_3aq $\xrightarrow{H_2O_2}$ $[Co(OH)(NH_3)_5]^{2+}$

$\xrightarrow{SO_3^{2-}}$ $[Co(SO_3)_2(NH_3)_4]^-$ $\xrightarrow{CN^-}$ $[Co(CN)(SO_3)(NH_3)_4]$

ii) $[Co(CN)(SO_3)(NH_3)_4]$ $\xrightarrow{H_3O^+}$ $[Co(CN)(NH_3)_4(H_2O)]^{2+}$

iii) $[Co(CN)(NH_3)_4(H_2O)]^{2+}$ $\xrightarrow[\text{charcoal}]{NH_3\text{aq}}$ $[CoCN(NH_3)_5]^{2+}$

Reaction i) giving an insoluble complex $[Co(CN)(SO_3)(NH_3)_4] \cdot 2\,H_2O$ was used with reference to an earlier work [76]. In reaction ii), the insoluble complex was dissolved in conc. H_2SO_4 at 0 °C, the solution was then poured into water, and conc. HCl was added to form the aqua complex $[Co(CN)(NH_3)_4(H_2O)]Cl_2$. In reaction iii), the aqua complex was allowed to react with aqueous NH_3, NH_4Cl and activated charcoal to produce $[CoCN(NH_3)_5]Cl_2$ (in analogy to the reaction in the familiar method for $[Co(NH_3)_6]Cl_3$).

Later, Siebert et al. [77] prepared *fac*-triamminetricyanocobalt(III) by the following reactions;

cis-$Na_2K_3[Co(CN)_4(SO_3)_2]$ $\xrightarrow[\text{evaporate}]{\text{aqua regia}}$

solid $\xrightarrow{\text{conc. HCl}}$ residue $\xrightarrow[\text{in autoclave}]{\text{conc. } NH_3}$ fac-$[Co(CN)_3(NH_3)_3] \cdot {}^1/_3\,H_2O$.

In 1951 Rây and Sarma [78] devised a method for obtaining the dicyanobis-(ethylenediamine) complex starting from bis(ethylenediamine)bis(thiosulfato) complexes;

$trans$- or cis-$[Co(S_2O_3)_2(en)_2]^+$ $\xrightarrow{CN^-}$

$[Co(CN)_2(en)_2]^+$ (thiosulfate salt)

The reaction gave a product regarded as *trans*-isomer. But, later, Chan and Tobe [71] assigned it to the *cis*-isomer on the basis of optical resolution. Ohkawa et al. [79] prepared the corresponding bis(propylenediamine) and bis(trimethylene-diamine) complexes, cis-$[Co(CN)_2(pn)_2]^+$ and cis-$[Co(CN)_2(tn)_2]^+$, by reference to Rây's method. In addition, they prepared a tetracyanoethylenediamine complex $[Co(CN)_4(en)]^-$ by the following reactions:

$trans$(Cl)-$[CoCl_2(NH_3)_2(en)]^+$ $\xrightarrow{S_2O_3^{2-}}$ $\xrightarrow{CN^-}$

$[Co(CN)_4(en)]^-$ (K or Na salt)

1.5.2 Charcoal Activation Method

Nagarajaiah et al. [80] examined the interaction of each of the ammine complexes such as $[Co(NH_3)_6]^{3+}$, $[Co(NH_3)_5(H_2O)]^{3+}$ and $[CoCl(NH_3)_5]^{2+}$ with an excess of cyanide in aqueous solution at room temperature, and they deduced the formation of pentacyano complexes. Block [81] attempted to prepare some of the amminecyano complexes by the action of CN^- ions on $[Co(NH_3)_6]^{3+}$, but the reaction gave an insoluble compound, $[Co(NH_3)_6][Co(CN)_6]$. However, Konya et al. [82] discovered that the reaction of a luteo-type complex with CN^- ions in a cold aqueous solution gave mixed cyano complexes in the presence of activated charcoal. The reaction was as follows:

$$[Co(NH_3)_6]^{3+} \xrightarrow[\substack{0 \sim 5\,°C \\ \text{charcoal}}]{CN^-} \begin{array}{l} [Co(CN)(NH_3)_5]^{2+} \\ [Co(CN)_2(NH_3)_4]^+ \ (trans\ and\ cis) \\ [Co(CN)_3(NH_3)_3] \ (mer) \end{array}$$

etc.

where the luteo complex was used as acetate because of its high solubility in water. A reaction mixture of the acetate and KCN in the mole ratio of 1:3, produced a mixture of some amminecyano complexes after a week in a refrigerator. The complexes were easily separated by ion-exchange chromatography (Dowex 50W-X8 in Li^+ form). In this method, the use of $[Co(en)_3]Cl_3$ or $[Co(dien)_2]Cl_3$ instead of $[Co(NH_3)_6](C_2H_3O_2)_3$ resulted in the isolation of trans- and cis-$[Co(CN)_2(en)_2]^+$ or mer-$[Co(CN)_3(dien)]$ (dien = diethylenetriamine).

The most striking characteristic of the mentioned method is the use of activated charcoal in cold medium. Under such a condition a luteo-type complex may be labilized by charcoal, and then equilibration is established among the possible products through ligand scrambling. In this sense the method for the preparation of mixed amminecyano complexes may be termed "charcoal activation method".

This method was successful in obtaining several stereoisomers of the $[Co(CN)_2-(R-pn)_2]^+$ complex [83]. A similar reaction between $[Cr(en)_3]^{3+}$ and CN^- resulted in the isolation of trans- and cis-$[Cr(CN)_2(en)_2]^+$ [84]. Recently, two new complexes trans- and cis-$[Co(CN)_4(NH_3)_2]^-$ and a known complex $[Co(CN)_5(NH_3)]^{2-}$ [85] were prepared from a reaction mixture of $[Co(NH_3)_6](C_2H_3O_2)_3$ and KCN in the mole ratio of 1:3.5 [86]. These complexes were effectively separated by Sephadex chromatography. Thus, the ammine-cyano series of complexes were completed; this is the first example for the complete series of mixed ligand complexes consisting of two kinds of unidentate ligand.

Table 1.1. Absorption Spectral Data for Ammine-Cyano Complexes ($\tilde{v}/10^3$ cm^{-1})

	$\tilde{v}$ (log ε)	$\tilde{v}$ (log ε)
$[Co(NH_3)_6]^{3+}$	21.0 (1.75)	29.4 (1.66)
$[Co(CN)(NH_3)_5]^{2+}$	22.7 (1.74)	30.6 (1.71)
trans-$[Co(CN)_2(NH_3)_4]^+$	ca. 21 (sh), 23.9 (1.82)	31.4 (1.83)
cis-$[Co(CN)_2(NH_3)_4]^+$	25.3 (1.84)	32.0 (1.77)
mer-$[Co(CN)_3(NH_3)_3]$	23.4 (sh), 26.8 (2.00)	33.3 (1.89)
fac-$[Co(CN)_3(NH_3)_3]$	26.4 (2.20)	34.5 (2.15)
trans-$[Co(CN)_4(NH_3)_2]^-$	24.2 (1.76), 31.0 (1.92)	ca. 36 (sh)
cis-$[Co(CN)_4(NH_3)_2]^-$	28.0 (2.00)	33.7 (1.90)
$[Co(CN)_5(NH_3)]^{2-}$	28.8 (2.30)	36.9 (2.04)
$[Co(CN)_6]^{3-}$	32.0 (2.29)	38.8 (2.12)

Table 1.1 gives the data on the first and second absorption bands for the entire set of complexes of the series which will be useful for obtaining information about the symmetry of chromophore and the electronic d—d transition.

Other methods for the preparation of the ammine-cyano complexes: Shibata et al. [12] obtained $[CoCN(NH_3)_5]^{2+}$ and *mer*-$[Co(CN)_3(NH_3)_3]$ from the reaction of KCN on $[Co(CO_3)_3]^{3-}$. In a similar manner, Kondo et al. [87] produced *fac*-$[Co(CN)_3(NH_3)_3]$. Muto et al. [88] synthesized *trans*-$[Co(CN)_2(en)_2]^+$ from a reaction mixture of *trans*-$[CoCl_2(en)_2]^+$ and KCN in DMSO. Maki and Sakuraba [89] obtained *cis*-$[Co(CN)_2(en)_2]^+$ by the reaction of AgCN with *trans*-$[CoCl_2(en)_2]^+$ in aqueous solution. Kawaguchi and Kawaguchi [90] prepared *trans*-$[Co(CN)_2(tn)_2]^+$ (tn = tri-methylenediamine) by the ligand substitution reaction of *trans*-$[CoCl_2(tn)_2]^+$ in methanol.

In 1968, Nishikawa et al. [91] extended the charcoal activation method to obtain the bis(acetylacetonato)dicyano complex, *cis*-$K[Co(CN)_2(acac)_2]$, and then prepared a triphenylphosphine-containing complex, $[Co(CN)_2(acac)(PPh_3)_2]$:

$$[Co(acac)_3] \xrightarrow[\substack{-5\sim0\ ^\circ C \\ charcoal}]{\substack{KCN \\ MeOH,}} \xrightarrow{Al_2O_3\ chromatography}$$

$$K[Co(CN)_2(acac)_2] \xrightarrow[boiling\ EtOH]{PPh_3} [Co(CN)_2(acac)(PPh_3)_2]$$

Kashiwabara et al. [92] communicated optical resolution of (2-aminoethyl)-n-butylphenylphosphine and the preparation of the bis(acetylacetonato)cobalt(III) complex containing the resolved aminophosphine. The resolved phosphine was allowed to react with tris(acetylacetonato)cobalt(III) in methanol solution in the presence of activated charcoal at room temperature. From the resulting solution a pair of diastereoisomers as shown in Fig. 1.6 was isolated by column chromatography using a SP-Sephadex C-25 ion exchanger and an aqueous solution of $Na_2[Sb_2(d\text{-tart})_2]$.

Kashiwabara et al. [93] reported the preparation of *cis*- and *trans*-bis(acetyl-acetonato)bis(dimethylphenylphosphine)cobalt(III) complexes, *cis*- and *trans*-$[Co(acac)_2(PMe_2Ph)_2]PF_6$. To this aim a mixture of $[Co(acac)_3]$, PMe_2Ph, and charcoal in ethanol—tetrahydrofuran (2:1) is stirred at room temperature for 12 h. The solution is then column-chromatographed with SP-Sephadex C-25. The geometrical

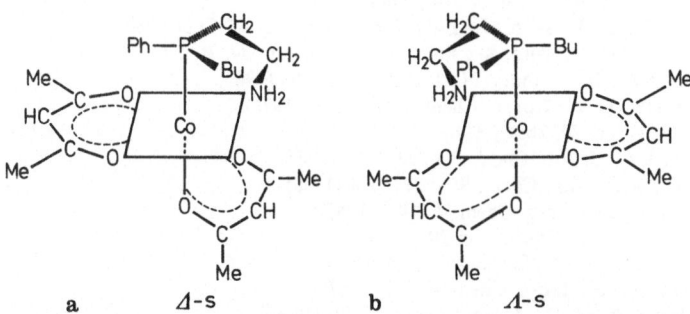

Fig. 1.6a and b. A pair of the diastereomers of the S-phosphine complex (from Ref. [92])

structures were determined by the ^{1}H and ^{13}C NMR spectra. The intensity of the first absorption band for the *trans*-isomer is 40 times as large as that of the *cis*-isomer.

Utilizing the facile substitution of a nitrite ion in $Na[Co(NO_2)_2(acac)_2]$, Hayakawa et al. [94] prepared the *cis*(NO_2, P)-$[Co(NO_2)(acac)_2(P)]$ type complexes (P = PBu_3, PBu_2Ph, PMe_2Ph and $PMePh_2$) by the reaction between the dinitro complex and trialkylphosphine or alkylarylphosphine in an organic medium. Triphenylphosphine did not yield the corresponding complex; the reaction of PMe_2Ph on $[Co(NO_2)_2$-$(acac)_2]^-$ gave *trans*(P, P)-$[Co(NO_2)_2(acac)(PMe_2Ph)_2]$.

1.6 References

1. Erdmann, O. L.: J. Prakt. Chem. *97*, 406 (1866)
2. Werner, A.: Z. Anorg. Chem. *8*, 174 (1895)
3. Jorgensen, S. M.: Z. Anorg. Chem. *17*, 475 (1898)
4. Duval, R.: J. Pure Appl. Chem. *13*, 468 (1938)
5. Sueda, H.: Bull. Chem. Soc. Jpn. *13*, 450 (1938)
6. Birk, E.: Z. Anorg. Chem. *175*, 409 (1928)
7. Tanito, Y., Saito, Y., Kuroya, H.: Bull. Chem. Soc. Jpn. *25*, 188 (1952)
8. Majumdar, A. K., Duval, C., Lecomte, J.: C.R. Acad. Sci. *247*, 302 (1958)
9. Mori, M., Shibata, M., Hirota, H., Masuno, K., Suzuki, Y.: Nippon Kagaku Zasshi *79*, 1251 (1958)
10. Penland, R. B., Lane, T. J., Quagliano, J. V.: J. Am. Chem. Soc. *78*, 887 (1956)
11. Maddock, A. G., Todesco, A. B. J. B.: J. Inorg. Nucl. Chem. *26*, 1535 (1964)
12. Shibata, M., Mori, M., Kyuno, E.: Inorg. Chem. *3*, 1573 (1964)
13. Hagel, R. B., Druding, L. F.: Inorg. Chem. *9*, 1496 (1970)
14. Cooley, W. E., Liu, C. F., Bailar, J. C.: J. Am. Chem. Soc. *81*, 4189 (1959)
15. Siebert, H.: Z. Anorg. Allg. Chem. *441*, 47 (1978)
16. Linhard, M., Siebert, H.: Z. Anorg. Allg. Chem. *364*, 24 (1969)
17. The X-ray studies refer to: Y. Komiyama, Bull. Chem. Soc. Jpn. *30*, 13 (1957)
18. Inorg. Synth. *2*, 216 (1946)
19. Sörensen, S. P. L.: Z. Anorg. Chem. *11*, 2 (1896)
20. Inorg. Synth. *1*, 37 (1939)
21. Dwyer, F. P., Reid, I. K., Garvan, F. L.: J. Am. Chem. Soc. *83*, 1285 (1961)
22. Hidaka, J., Shimura, Y., Tsuchida, R.: Bull. Chem. Soc. Jpn. *35*, 567 (1962)
23. Inorg. Synth. *4*, 172 (1953)
24. Hund, J. P., Rutenberg, A. C., Taube, H.: J. Am. Chem. Soc. *74*, 268 (1952)
25. Fujita, J., Yasui, T., Shimura, Y.: Bull. Chem. Soc. Jpn. *38*, 654 (1965)
26. Werner, A., Humphrey, E.: Ber. *34*, 1720 (1901); Inorg. Synth. *4*, 177 (1953)
27. Legg, J. I., Cooke, D. W.: Inorg. Chem. *5*, 594 (1966)
28. Mizukami, F., Ito, H., Fujita, J., Saito, K.: Bull. Chem. Soc. Jpn. *45*, 2129 (1972)
29. Mizukami, F., Ito, H., Fujita, J., Saito, K.: Bull. Chem. Soc. Jpn. *44*, 3051 (1971)
30. Boucher, L. J., Bailar, J. C., Jr.: J. Inorg. Nucl. Chem. *27*, 1093 (1963)
31. Brubaker, G. R., Schaefer, D. P.: Coord. Chem. Rev. *7*, 161 (1971)
32. Basolo, F.: J. Am. Chem. Soc. *70*, 2634 (1948)
33. Das Sarma, B., Bailar, J. C.: J. Am. Chem. Soc. *77*, 5480 (1955)
34. Selbin, J., Bailar, J. C., Jr.: J. Am. Chem. Soc. *82*, 1524 (1960)
35. Buckingham, D. A., Jones, D.: Inorg. Chem. *4*, 1387 (1965)
36. Gillard, R. D., Wilkinson, G.: J. Chem. Soc. *1963*, 3193
37. Buckingham, D. A., Marzilli, P. A., Sargeson, A. M.: Inorg. Chem. *6*, 1032 (1967)
38. Sargeson, A. M., Searle, G. H.: Inorg. Chem. *6*, 787 (1967)
39. Hamilton, H. G., Alexander, M. D.: Inorg. Chem. *5*, 2060 (1966)
40. Bosnich, B., Gillard, R. D., McKenzie, E. D., Webb, G. A.: J. Chem. Soc. A *1966*, 1331

41. Alexander, M. D., Hamilton, H. G.: Inorg. Chem. 8, 2131 (1969)
42. Brubaker, G. R., Schaefer, D. P.: Inorg. Chem. 10, 968 (1971)
43. Goto, M., Saburi, M., Yoshikawa, S.: Inorg. Chem. 8, 358 (1969)
44. Saburi, M., Yoshikawa, S.: Bull. Chem. Soc. Jpn. 45, 806 (1972)
45. Saburi, M., Sawai, T., Yoshikawa, S.: Bull. Chem. Soc. Jpn. 45, 1086 (1972)
46. Saburi, M., Yoshikawa, S.: Bull. Chem. Soc. Jpn. 45, 1443 (1972)
47. Goto, M., Matsushita, H., Saburi, M., Yoshikawa, S.: Inorg. Chem. 12, 1498 (1973)
48. Goto, M., Makino, T., Saburi, M., Yoshikawa, S.: Bull. Chem. Soc. Jpn. 49, 1879 (1976)
49. Yoshikawa, S., Saburi, M., Yamaguchi, M.: Pure & Appl. Chem. 50, 915 (1978)
50. Worrell, J. H., Busch, D. H.: Inorg. Chem. 8, 1563 (1969)
51. Worrell, J. H., Busch, D. H.: Inorg. Chem. 8, 1572 (1969)
52. Bosnich, B., Kneen, W. R., Phillip, A. T.: Inorg. Chem. 8, 2567 (1969)
53. Schrauzer, G. N., Kohnle, J.: Chem. Ber. 97, 3056 (1964)
54. Schrauzer, G. N., Windgassen, R. J., Kohnle, J.: Chem. Ber. 98, 3324 (1965)
55. Schrauzer, G. N., Windgassen, R. J.: Chem. Ber. 99, 602 (1966)
56. Inorg. Synth. 11, 61 (1968)
57. Sadasivan, N., Kernohan, J. A., Endicott, J. F.: Inorg. Chem. 6, 770 (1967)
58. Curtis, N. F.: J. Chem. Soc. 1960, 4409
59. Blight, M. M., Curtis, N. F.: J. Chem. Soc. 1962, 3016
60. Curtis, N. F.: J. Chem. Soc. 1964, 2644
61. Curtis, N. F., Curtis, Y. M., Powell, H. K.: J. Chem. Soc. 1966, 1015
62. Sadasivan, N., Endicott, J. F.: J. Am. Chem. Soc. 88, 5468 (1966)
63. Travis, K., Busch, D. H.: Inorg. Chem. 13, 2591 (1974)
64. Hathaway, B. J., Hofah, D. G., Underhill, A. E.: J. Chem. Soc. 1962, 2444
65. Bailar, J. C., Jr. (ed.): The chemistry of the coordination compounds Chapt. 6. Reinhold Publishing Coorporation 1956
66. Ogino, H., Fujita, J.: Chem. Lett. 1973, 517
67. Ogino, H., Fujita, J.: Bull. Chem. Soc. Jpn. 48, 1836 (1975)
68. Ogino, H.: Bull. Chem. Soc. Jpn. 50, 2459 (1977)
69. Fujita, J., Ogino, H.: Chem. Lett. 1974, 57
70. Chan, S. C., Tobe, M. I.: J. Chem. Soc. 1968, 966
71. Chan, S. C.: J. Chem. Soc. 1964, 2716
72. Ohkawa, K., Hidaka, J., Shimura, Y.: Bull. Chem. Soc. Jpn. 39, 1715 (1966)
73. Werner, A., Grüger, H.: Z. Anorg. Chem. 16, 398 (1898)
74. Scott, K. I.: J. C. S. Dalton 1974, 1486
75. Siebert, H.: Z. Anorg. Allg. Chem. 327, 63 (1964)
76. Hofmann, K. A., Reinsch, S.: Z. Anorg. Allg. Chem. 16, 277 (1898)
77. Siebert, H., Siebert, C., Wieghardt, K.: Z. Anorg. Allg. Chem. 380, 30 (1971)
78. Rây, P. R., Sarma, B.: J. Indian Chem. Soc. 28, 59 (1951)
79. Ohkawa, K., Fujita, J., Shimura, Y.: Bull. Chem. Soc. Jpn. 38, 66 (1965)
80. Nagarajaiah, H. S., Sharpe, A. G., Wakefield, D. B.: Proc. Chem. Soc. 1959, 385
81. Block, B. P.: J. Inorg. Nucl. Chem. 14, 294 (1960)
82. Konya, K., Nishikawa, H., Shibata, M.: Inorg. Chem. 7, 1165 (1968)
83. Kashiwabara, K., Yamanaka, T., Saito, K., Komatsu, N., Hamada, N., Nishikawa, H., Shibata, M.: Bull. Chem. Soc. Jpn. 48, 3631 (1975)
84. Kaizaki, S., Hidaka, J., Shimura, Y.: Bull. Chem. Soc. Jpn. 48, 902 (1975)
85. Cambi, L., Daglia, E.: Gazz. Chim. Ital. 88, 691 (1964)
86. Fujinami, S., Nakada, T., Shibata, M.: unpublished
87. Kondo, Y., Nakahara, M., Kataoka, H., Yamamoto, H.: Nippon Kagaku-Zasshi 92, 272 (1971)
88. Muto, M., Baba, T., Yoneda, H.: Bull. Chem. Soc. Jpn. 41, 2918 (1968)
89. Maki, N., Sakuraba, S.: Bull. Chem. Soc. Jpn. 42, 1908 (1969)
90. Kawaguchi, H., Kawaguchi, S.: Bull. Chem. Soc. Jpn. 46, 3453 (1973)
91. Nishikawa, H., Konya, K., Shibata, M.: Bull. Chem. Soc. Jpn. 41, 1492 (1968)
92. Kashiwabara, K., Kinoshita, I., Fujita, J.: Chem. Lett. 1978, 673
93. Kashiwabara, K., Katoh, K., Fujita, J., Nishikawa, H., Shibata, M.: Chem. Lett. 1981, 575
94. Hayakawa, S., Nishikawa, H., Kashiwabara, K., Shibata, M.: Bull. Chem. Soc. Jpn. 54, 3593 (1981)

2 Versatile Uses of Tricarbonatocobaltate(III) as Starting Material

2.1 Historical

From the absorption spectra of the various green solutions produced by the reaction of hydrogen peroxide on a cobaltous salt in the presence of potassium (or sodium) bicarbonate, acetate, tartrate, citrate or oxalate, Durrant [2,3] concluded that the green color depended on complex ions with the $>Co \cdot O \cdot Co<$ nucleus, and that the differences in tint of green and in absorption spectra depended probably on the various association of carbonyl groups attached to the nucleus. On account of his detailed study, the reaction discovered by Field [1] has been called "Field-Durrant reaction" [2-8].

The green solution of carbonatocobaltate(III) was prepared from a cobalt(III) coordination compound by McCutcheon and Schule [9]; a solution of tetraammine-carbonatocobalt(III) sulfate was added to a hot aqueous solution of potassium bicarbonate and potassium persulfate. The mixture was heated in a steam-bath until its red color changed to dark green. When a hot aqueous solution of hexa-amminecobalt(III) nitrate was added to the above solution, a grayish-green compound precipitated. Its analysis matched the formula $[Co(NH_3)_6][Co(CO_3)_3]$.

Mori and Shibata [10] modified Laitinen's method [8] for the determination of cobalt. With much ammonium chloride a green solution of the Field-Durrant reaction changes from green to blue-violet, then to red-violet upon continued heating. The formation of tetraamminecarbonatocobalt(III) was confirmed after completion of the reaction. Moreover, oxalic acid reacts with the green solution to form tris(oxalato)cobaltate(III) ion with the evolution of carbon dioxide. The absorption spectrum of the green carbonato solution is similar to that of tris(oxalato)cobaltate(III), so that the carbonato-complex anion must be the tricarbonatocobaltate(III). The isolation of its potassium salt, $K_3[Co(CO_3)_3] \cdot 3 H_2O$, was reported soon after [11]. This success was due to the way of preparing the green solution in a preparative scale. Another method for preparing cobalt(III) complexes starts from a freshly prepared green tricarbonatocobaltate(III) solution.

Some years later, Bauer and Drinkard [12] prepared the hardly soluble sodium salt of tricarbonatocobaltate(III), $Na_3[Co(CO_3)_3] \cdot 3 H_2O$, in an analogous manner. They used solid sodium salt as the starting material for preparing certain cobalt(III) coordination compounds.

The carbonato complexes of cobalt(III) have been comprehensively reviewed by MacColl [13] and the chemistry of metal carbonato complexes by Krishnamurty et al. [14].

2.2 Tricarbonatocobaltate(III) Anion

2.2.1 Preparation

The tricarbonatocobaltate(III) has been more interesting to preparative chemists than to analytical chemists. The complex has been used as starting material for various cobalt(III) complexes. It was prepared by three procedures:

Procedure 1 [10, 11, 15]. $KHCO_3$ (35 g, 0.35 mol) is introduced into 35 cm^3 water, and the resulting slurry is kept cool in a freezing mixture of ice and sodium chloride. Separately, 12 g of $CoCl_2 \cdot 6 H_2O$ (0.05 mol) in 12 cm^3 of hot water is mixed with 20 cm^3 of 30% H_2O_2 in ice. The $Co^{2+} - H_2O_2$ mixture is added dropwise to the quenched slurry with constant stirring, and the resulting green solution is then quickly filtered by suction. The clear filtrate is again cooled in an ice-bath and used as starting material. A new batch should be prepared for each run of experiments. In place of cobalt chloride, the use of the nitrate is convenient because it is more soluble in water.

Procedure 2 [12, 16]. A solution of 29.1 g (0.10 mol) of $Co(NO_3)_2 \cdot 6 H_2O$ and 10 cm^3 of 30% H_2O_2 in 50 cm^3 of water is added dropwise with stirring to an ice-cold (0 °C) slurry of 42.0 g (0.5 mol) of $NaHCO_3$ in 50 cm^3 of water. After stirring for one hour at 0 °C, the olive-green products is filtered, washed with three 10 cm^3 portion of cold water, and then with absolute ethanol and dry ether. The product is further dried over phosphorus(V) oxide. The product must be thoroughly dry when it is stored. Small amounts of water decompose the compound into a black solid. The yield is 26.7 g (74%).

Procedure 3 [17]. Cobalt(II) chloride hexahydrate (23.8 g, 0.10 mol) is dissolved in 450 cm^3 of water, and 50 cm^3 of 30% H_2O_2 is added. Under vigorous shaking the solution is slowly poured onto 100 g (1.0 mol) of $KHCO_3$ [1]. The resulting dark green to black solution containing (probably among other ions) the tricarbonato-cobaltate(III) is kept for 5 minutes, to allow the excess peroxide to decompose.

Advantages and disadvantages of using solutions or solids will be mentioned later.

2.2.2 Properties

There is a striking solubility difference between the potassium and sodium salts of tricarbonatocobaltate(III). The concentration of the solution prepared according to Procedure 1 can be made as high as 1 mol/dm^3 with respect to Co(III); however the potassium salt can be isolated only with simultaneous decomposition to hydrous-oxide-like substances. In contrast, the sodium salt obtained by Procedure 2 is very stable when it is completely dried. The salt is insoluble in water but soluble in water containing sodium bicarbonate.

---

[1] The order of addition is important. This procedure affords a solution of high cobalt concentration (approximately 0.2 mol/dm^3) which decomposes the excess hydrogen peroxide at a desirable rate. Substitution of sodium hydrogen carbonate for the potassium salt makes it impossible to have such a high concentration of cobalt in solution.

The anhydrous salt $[Co(NH_3)_6][Co(CO_3)_3]$ made by the method of McCutcheon and Schule [9] or more simply by adding $[Co(NH_5)_6]X_3$ to a green solution prepared by Procedure 1, is much more stable when it is dry. The compound is soluble to the extent of only *ca.* 0.04 g per 100 cm^3 of water at room temperature, the solution being quite unstable. However, the compound is ten times more soluble in an aqueous solution of potassium or sodium bicarbonate. Baur and Bricker [18] used this compound as a reagent for titrimetric oxidation; the compound is a very weak oxidant in bicarbonate media, but in an acid solution the cobalt(III) in the $[Co(CO_3)_3]^{3-}$ portion is released and reacts quantitatively with reductants such as iron(II), vanadium(IV) and cerium(III).

The absorption spectrum of the Field-Durrant solution with potassium bicarbonate is characteristic of the cobalt(III) complex; [10] there are two d—d absorption bands and one charge-transfer (CT) band, and the pattern of the spectrum is similar to that of the familiar tris(oxalato)cobaltate(III) anion except for some bathochromic shifts of the carbonato complex. Regarding the d—d bands, conformity to Beer's law is confirmed with *ca.* $10^{-3} \sim 10^{-4}$ mol/dm^3 solutions using 1-cm absorption cell. Data on the solution spectrum with the values of the parameters Δ and B [19] are given in Table 2.1 [10].

Table 2.1. Absorption Spectral Data on Tricarbonatocobaltate(III) and Related Complexes ($\tilde{v}/10^3$ cm^{-1})

Complex	$\tilde{v}_I$(log ε)	$\tilde{v}_{II}$(log ε)	Δ	B (cm^{-1})
$[Co(CO_3)_3]^{3-}$	15.5 (2.19)	22.7 (2.22)	16.1	550
$[Co(ox)_3]^{3-}$	16.5 (2.17)	23.6 (2.30)	17.2	530
$[Co(mal)_3]^{3-}$	16.4 (2.17)	23.8 (2.10)	17.1	560
$[Co(H_2O)_6]^{3+}$	16.6 (1.6)	24.9 (1.7)	17.1	650
$[Co(NO_3)_6]^{3-}$ [a]				
in acetonitrile	15.2 (1.92)	22.5 (2.00)	16.2	466

a Khalil, M. I., Logan, N., Harris, A. D.: J. C. S. Dalton, *1980*, 314

The UV maximum is recommended by Telep and Boltz [20] for the spectrophotometric determination of small amounts of cobalt in various samples because of the conformity to Beer's law over a wide range of concentrations.

After IR measurement with the compound of incorrect formula $Co[Co(CO_3)_3]$ [21], a more precise measurement was carried out by Gatehouse et al. [22] with Nujol. Several results are cited in Table 2.2, [23, 24] Lapscombe [25] also examined the infrared spectrum of the compound in the KBr disc and concluded that the compound contained a unidentate carbonate group, supporting the formula $[Co(CO_3)_3(H_2O)_3]^{3-}$.

Prior to these studies, Nakamoto et al. [26] proposed a convenient method of distinguishing between unidentate carbonate ion and bidentate one; in the spectra between 1250 and 1650 cm^{-1}, the band separation between two observed bands is greater in a complex with bidentate carbonate than in that with unidentate carbonate. From this method, Gillard et al. [27] judged the coordination of carbonate groups in some tricarbonatocobaltate(III) salts employing KBr disc and D_2O solution. Their results are listed in Table 2.3. The compound $[Co(NH_3)_6][Co(CO_3)_3]$ with bands at 1585 and 1280 cm^{-1} contains bidentate CO_3^{2-} ions, while the compound

Table 2.2. IR Data on the Carbonato Complexes [23,24] (ν/cm^{-1})

Species (C$_{2v}$)	$\nu_1(A_1)$	$\nu_2(A_1)$	$\nu_3(A_1)$	$\nu_5(B_2)$	$\nu_8(B_1)$
Calc. Frequency	1595	1038	771	1282	—
Assignment	ν(C—O$_{II}$)	ν(C—O$_{I}$)	Ring def. $+\nu$(Co—O$_{I}$)	ν(C—O$_{I}$) $+\delta$(O$_{I}$CO$_{II}$)	π
[Co(CO$_3$)(en)$_2$]Cl	1577s	1059w 1035w	754m	1281s 1272s	754m
[Co(CO$_3$)(NH$_3$)$_4$]Cl[24]	1593	1030	760	1265	834
K$_3$[Co(CO$_3$)$_3$] · 3 H$_2$O	1527s	1080w 1037m	809m	1330s	851m
[Co(NH$_3$)$_6$] [Co(CO$_3$)$_3$]	1523s	1073w 1031w	738m	1285s	889w

Species (C$_{2v}$)	$\nu_1(A_1)$	$\nu_2(A_1)$	$\nu_3(A_1)$	$\nu_5(B_2)$	$\nu_8(B_1)$
Calc. Frequency	1376	1069	772	1482	—
Assignment	ν(C—O$_{II}$) $+\nu$(C—O$_{I}$)	ν(C—O$_{I}$) $+\nu$(C—O$_{II}$)	δ(O$_{II}$CO$_{II}$)	ν(C—O$_{II}$)	π
[CoCO$_3$(NH$_3$)$_5$]Cl	1297s	1057m 1043m	738m	1493s	873s 848s
[CoCO$_3$(NH$_3$)$_5$]Br[24]	1373	1070	756	1453	850

Table 2.3. IR Results on Carbonato-Complexes [27] (ν_3/cm^{-1} of D$_{3h}$ Carbonate)

Compound	KBr disc	D$_2$O solution
Free CO$_3^{2-}$	1415 [26]	
KHCO$_3$	1625, 1400	
[Co(NH$_3$)$_6$]Cl · CO$_3$	1390 ~ 1370 [26]	
[Co(CO$_3$)(NH$_3$)$_5$]Br	1450, 1370 [26]	
[Co(CO$_3$)(NH$_3$)$_5$]NO$_3$	1470, 1360	1460, 1360
[Co(CO$_3$)(NH$_3$)$_4$]NO$_3$	1600, 1282	1610, 1355
[Co(CO$_3$)(en)$_2$]Br	1575, 1278	1610, 1350
[Co(NH$_3$)$_6$][Co(CO$_3$)$_3$]	1585, 1280	insoluble
(+)[Co(en)$_3$](−)[Co(CO$_3$)$_3$]	1600, 1300	insoluble
K$_3$[Co(CO$_3$)$_3$] · 3 H$_2$O	1600, 1495, 1335	1600, 1470, 1345

K$_3$[Co(CO$_3$)$_3$] · 3 H$_2$O with two bands at 1495 and 1335 cm^{-1}, and a weaker band at 1600 cm^{-1} contains mainly unidentate CO$_3^{2-}$ ions.

For the optical resolution [27], the less soluble diastereoisomer (+)$_{589}$[Co(en)$_3$]-(−)$_{589}$[Co(CO$_3$)$_3$] was precipitated by mixing cold (0 °C) saturated solutions of K$_3$[Co(CO$_3$)$_3$] · 3 H$_2$O and (+)[Co(en)$_3$]Cl$_3$ and maintaining the mixture at 0 °C for a few hours. The diastereoisomer exhibited very weak Cotton effects when the circular dichroism spectrum was measured in the KBr disc, while the filtrate showed no optical activity. Attempts to dissolve the diastereoisomer gave optically inactive solution. From these facts, the workers concluded that the chelated ion $[Co(CO_3)_3]^{3-}$ exists in the hexaamminecobalt(III) or the tris(ethylenediamine)-cobalt(III) salt, while in solution of the potassium salt, the concentration of carbonato chelates will be low and successive equilibria are established;

$$[Co(CO_3)_3]^{3-} \rightleftarrows [Co(CO_3)_2(HCO_3)(OH)]^{3-} \rightleftarrows$$
$$\rightleftarrows [Co(CO_3)(HCO_3)_2(OH)_2]^{3-} \rightleftarrows [Co(HCO_3)_3(OH)_3]^{3-}$$

The procedure used for this resolution was applied to kinetically labile complexes of the type $[M(ox)_3]^{3-}$ (M = V, Cr, Mn, Fe, or Co), using the kinetically inert complexes, $(+)_{589}[Co(en)_3]^{3+}$ and $(-)_{589}[Rh(en)_3]^{3+}$, as the resolving agents [28]. The CD measurements were made on the less soluble diastereoisomers in KBr, KCl, or CsCl discs.

Segupta and Nandi have studied related complex carbonates of chromium(III) [29] and iron(III) [30]. The absorption spectrum of $Cr(NO_3)_3 \cdot 6 H_2O$ in $KHCO_3$ solution (35%) showed two absorption maxima at 17240 cm^{-1} and 23250 cm^{-1}, suggesting the formation of $[Cr(CO_3)_3]^{3-}$, but the solid compound isolated from a Cr(III)-alum-$KHCO_3$ solution by adding ethanol was $K_7[Cr_4(OH)_9(CO_3)_5] \cdot 6 H_2O$. They considered this compound as the hydrolysis product of hypothetical $[Cr(CO_3)_3]^{3-}$, namely,

$$8 K_3[Cr(CO_3)_3] + 14 H_2O \rightarrow 2 K_7[Cr_4(OH)_9(CO_3)_3] +$$
$$+ 4 CO_2 + 10 KHCO_3$$

In the study on the iron(III) complexes, no tricarbonatoferrate(III) ion was found in $Fe(NO_3)_3$aq-$KHCO_3$ solution and the compounds isolated contained $[Fe_2(OH)_4(CO_3)_3]^{4-}$ and $[Fe_3(OH)_4(CO_3)_6]^{7-}$

When these results are considered, the fact that Procedure 1 sometimes results in a turbid green solution may be due to hydrolysis of the carbonatocobaltate(III) anion.

2.3 Systematic Synthesis of Complexes

The use of "the green solution" prepared according to Procedure 1 enables us to displace the ligating CO_3^{2-} ions successively by a ligand of a higher ligand field; the stepwise substitution gives a dicarbonato-complex such as $[Co(CO_3)_2a_2]$ or $[Co(CO_3)_2(AA)]$, then a monocarbonato-complex such as $[Co(CO_3)a_4]$ or $[Co(CO_3)-(AA)_2]$, and finally a complex such as $[Co\, a_6]$ or $[Co(AA)_3]$. Thus, complexes can be synthesized systematically by controlling the kind and amount of the reagent for ligand, temperature, etc.

2.3.1 Ammine-Carbonato and Carbonato-Diamine Series

The "the green solution" reacts with commercial ammonium carbonate on a water-bath to form diamminedicarbonatocobaltate(III) anion, and its potassium salt cis-K[Co(CO$_3$)$_2$(NH$_3$)$_2$] · H$_2$O has been isolated [11, 31]. By using aqueous ammonia instead of ammonium carbonate the reaction proceeds at room temperature.[2] The reaction of "the green solution" with aqueous ammonia on a water-bath gives the familiar tetraamminecarbonatocobalt(III) salt, [CoCO$_3$(NH$_3$)$_4$]X. With activated charcoal the result is the hexaamminecobalt(III) salt, [Co(NH$_3$)$_6$]X$_3$.

Similarly, the corresponding carbonatoethylenediamine complexes have been synthesized [31, 32]. Originally [32] solid material prepared by saturating ethylenediamine hydrate with carbon dioxide was used in order to realize a mild substitution at the first-step, but we now recommend ethylenediamine instead of such salt-like material, because of convenience in procedure and a good yield of K[Co(CO$_3$)$_2$en] · H$_2$O.[3] Other familiar complexes, [CoCO$_3$(en)$_2$]X and [Co(en)$_3$]X$_3$, are also accessible.

Another series starts with 2,2'-bipyridine or 1,10-phenanthroline [34]. "The green solution" mixed with the ligand in ethanol at room temperature yields K[Co(CO$_3$)$_2$-(bpy)] · 2 H$_2$O or K[Co(CO$_3$)$_2$(phen)] · H$_2$O. The other complexes, [Co(CO$_3$)(bpy)$_2$]X [35], [Co(CO$_3$)(phen)$_2$]X [35], [Co(bpy)$_3$]X$_3$ and [Co(phen)$_3$]X$_3$ are obtained by the reaction of "the green solution" with the ligand in stoichiometric amounts, but activated charcoal is necessary to prepare the tris-type complexes.

2.3.2 Ammine-Oxalato and Diamine-Oxalato Series

The addition of calculated amounts of solid oxalic acid to a solution of tetra-amminecarbonato complex, [Co(CO$_3$)(NH$_3$)$_4$]X, yields the corresponding oxalato complex, [Co(ox)(NH$_3$)$_4$]X [36]. With solid oxalic acid, "the green solution" produces tris(oxalato)cobaltate(III), and the familiar potassium salt, K$_3$[Co(ox)$_3$] · 3 H$_2$O, is obtainable by acidifying with acetic acid. When solid ammonium oxalate is added to "the green solution", the carbonate ligands are substituted with both oxalate ion and ammonia, giving the potassium salt of diamminecarbonatooxalatocobalt-ate(III), K[Co(CO$_3$)(ox)(NH$_3$)$_2$] · H$_2$O [31, 32]. The addition of oxalic acid to a solution of this compound and the subsequent addition of ethanolic ammonium chloride result in the precipitation of cis-NH$_4$[Co(ox)$_2$(NH$_3$)$_2$] · H$_2$O, which is identical to that obtained from [CoCl$_2$(NH$_3$)$_2$(H$_2$O)$_2$]Cl with oxalic acid in aqueous solution [37].

Similarly, solid ethylenediammonium oxalate from the neutralization of the diamine with the acid form the carbonatoethylenediamineoxalatocobaltate(III) anion, the potassium salt K[CoCO$_3$(ox)(en)] · H$_2$O being isolated [31, 32]. The addition of oxalic

---

[2] Conc. aqueous ammonia (7 cm^3) is added to "the green solution" (Co(NO$_3$)$_2$ · 6 H$_2$O 14.5 g scale) and the mixture is stirred at room temperature until the color becomes inky blue. The resulting solution is rapidly cooled with ice, then ethanol (ca. 50 cm^3) is added. While standing in the cold an aqueous layer is formed and decanted. The addition of ethanol (ca. 30 cm^3) to the residual blue oil is repeated until the decanted solution becomes clear. From the final oily material the complex crystallizes on ice in about 10 min. The yield is ca. 6 g.

[3] K[Co(CO$_3$)$_2$(en)] · H$_2$O is synthesized in a similar manner as K[Co(CO$_3$)$_2$(NH$_3$)$_2$] · H$_2$O. The yield is 5 g. Rowan et al. [33] synthesized the same dicarbonatoethylenediamine complex from "the green solution" with limiting amounts of en · 2 HClO$_4$.

acid to this compound in water yields the ethylenediaminebis(oxalato) complex, K[Co(ox)$_2$(en)] · H$_2$O, prepared by Dwyer et al. [38] from a mixed solution of cobalt(II) acetate, oxalic acid and ethylenediammonium dichloride by PbO$_2$ oxidation.

2.3.3 Carbonato Complexes Mixed with an Aniono Ligand with Higher Ligand Field

Nitrite ions tend to form hexanitrocobaltate(III) anion with "the green solution". However, under ice cooling, stoichiometric amounts of potassium nitrite react with "the green solution" to yield the dicarbonatodinitrocobaltate(III) complex, cis-K$_3$[Co(CO$_3$)$_2$(NO$_2$)$_2$] · H$_2$O, and the carbonatotetranitrocobaltate(III) complex, K$_3$[CoCO$_3$(NO$_2$)$_4$] · H$_2$O [39, 40]. The addition of potassium nitrite and acetylacetone to "the green solution" results in the acetylacetonatocarbonatodinitro complex cis-K$_2$[Co(CO$_3$)(NO$_2$)$_2$(acac)] · H$_2$O.

Golovnya et al. [41] isolated K$_3$[Co(NO$_2$)$_2$(CO$_3$)$_2$] · 2 H$_2$O from the reaction mixture of "the green solution" and saturated aqueous potassium nitrite. Furthermore, they isolated a binuclear complex, K$_8$[Co$_2$(NO$_2$)$_8$(CO$_3$)$_3$] · 2 H$_2$O, from the reaction mixture of "the green solution" and saturated KNO$_2$ solution after 8 ~ 10 months [42].

The reaction of "the green solution" with potassium cyanide in stoichiometric amount at room temperature gives rise to dicarbonatodicyanocobaltate(III), which has been isolated as the tris(ethylenediamine)cobalt(III) salt, (+)$_{589}$[Co(en)$_3$] · cis-[Co(CN)$_2$(CO$_3$)$_2$] · 2 H$_2$O [43]. The addition of oxalic acid to the solution containing [Co(CN)$_2$(CO$_3$)$_2$]$^{3-}$ derives cis-[Co(CN)$_2$(CO$_3$)(ox)]$^{3-}$ or cis-[Co(CN)$_2$(ox)$_2$]$^{3-}$ depending upon the quantity of the acid used. Either complex has been isolated as the tris-(ethylenediamine)cobalt(III) salt, (+)$_{589}$[Co(en)$_3$] · cis-[Co(CN)$_2$(CO$_3$)(ox)] and (+)$_{589}$[Co(en)$_3$] · cis-[Co(CN)$_2$(ox)$_2$] [44].

In the described syntheses "the green solution" as the starting material has the following characteristics:

i)   The two CO$_3^{2-}$ ions in the [Co(CO$_3$)$_3$]$^{3-}$ anion are replaced stepwise by other ligands, giving dicarbonato- then monocarbonato-complex species, while the last CO$_3^{2-}$ ion is hard to be replaced by a ligand in the presence of excess bicarbonate ions. This enables us to synthesize various complexes.

ii) The substitution of a base such as ammonia or a diamine for all the CO$_3^{2-}$ ligands is effected by activated charcoal.

iii) The principal advantage of "the green solution" is to permit the syntheses of dicarbonato-complexes which have never been obtained by the so far known preparative methods.

iv) As for the [Co(CO$_3$)$_2$a$_2$]-type complexes, their geometrical structure is established to be cis with respect to the unidentate ligand a.[4] The absorption spectral data of the dicarbonato-complexes and some related complexes are summarized in Table 2.4.

v)   The substitution reactions occur almost stoichiometrically at mild conditions.

---

[4] For the [Co(CO$_3$)$_2$(NH$_3$)$_2$]$^-$ complex, the authors assumed a blue variety as cis- and a violet one as trans-isomer. Later, the formation of the violet variety was denied by Hyodo and Archer [45], who regarded it as a mixture of various species.

vi) A disadvantage is that some salts such as $KHCO_3$, KCl (or KNO_3), etc. remaining in the resulting reaction mixture sometimes interfere with the crystallization of a desired complex.

Table 2.4. Absorption Spectral Data on Dicarbonato-Type Complexes ($\tilde{v}/10^3$ cm^{-1})

Complex	$\tilde{v}_I$ (log ε)	$\tilde{v}_{II}$ (log ε)
cis-[Co(CO$_3$)$_2$(NH$_3$)$_2$]$^-$	17.40 (2.14)	25.60 (2.40)
[Co(CO$_3$)$_2$en]$^-$	17.53 (2.17)	25.63 (2.33)
[Co(CO$_3$)$_2$(bpy)]	17.80 (2.18)	ca. 25.6 sh (ca. 2.30)
[Co(CO$_3$)$_2$(phen)]$^-$	17.60 (2.18)	ca. 25.4 sh (ca. 2.40)
cis-[CoCO$_3$(ox)(NH$_3$)$_2$]$^-$	17.70 (2.09)	25.73 (2.18)
[CoCO$_3$(ox)(en)]$^-$	17.83 (2.17)	26.16 (2.30)
cis-[Co(CO$_3$)$_2$(NO$_2$)$_2$]$^{3-}$	18.26 (2.34)	—
cis-[Co(NO$_2$)$_2$(CO$_3$)(ox)]$^{3-}$	18.66 (2.26)	—
cis-[Co(NO$_2$)$_2$(CO$_3$)(acac)]$^{2-}$	18.70 (2.18)	—
cis-[Co(CN)$_2$(CO$_3$)$_2$]$^{3-}$	18.30 (1.96), 22.5 (2.03)	27.50 (2.20)
cis-[Co(CN)$_2$(CO$_3$)(ox)]$^{3-}$	18.70 (1.90), 23.00 (2.07)	28.10 (2.25)

2.4 General Syntheses of Tris(bidentate) Complexes

By far the tris-type complexes such as tris(diamine)cobalt(III), tris(dicarboxylato)-cobaltate(III), and tris(β-ketoacidato)cobalt(III) are best prepared by oxidizing the cobalt(II) ion in the presence of the ligand in basic form. However, Bauer and Drinkard [12] utilized sodium tricarbonatocobaltate(III) trihydrate in a general method to prepare:

> Tris(ethylenediamine)cobalt(III) chloride
> Tris(1,3-diamino-2-propanol)cobalt(III) nitrate
> Tris(acetylacetonato)cobalt(III)
> Tris(benzoylacetonato)cobalt(III)
> Tris(o-aminophenolato)cobalt(III) 1.5-hydrate[5]
> Sodium tris(mercaptoacetato)cobaltate(III) 6-hydrate[6]
> Tris(ethylenediamine)cobalt(III) tris(salicylato)cobaltate(III)
> Sodium tris(ethylenedimercapto)cobaltate(III) 2-hydrate

---

[5] Preparation: A slurry of Na$_3$[Co(CO$_3$)$_3$] · 3 H$_2$O (3.6 g, 0.01 mol) in ethanol (30 cm^3) is added to o-aminophenol hydrochloride (4.4 g, 0.03 mol) and the mixture is refluxed for an hour. After filtration the solution is concentrated to precipitate a brown product. It is washed with cold water, and recrystallized from ethanol-water. Yield, 3.4 g (88%).

[6] Preparation: An aqueous solution of mercaptoacetic acid (2.8 g, 0.03 mol in 50 cm^3) is added to the starting material (3.6 g, 0.01 mol) while stirring. The mixture is warmed on a steam-bath for 30 min. The dark green solid separates out upon cooling and is washed with acetone. Yield, 3.4 g (85%).

33

Bauer and Drinkard described several features of tricarbonatocobaltate(III) sodium as an intermediate [12]:

i) Because the source of cobalt(III) ion obviates the need for an oxidant, the reaction of compounds such as *o*-aminophenol, mercaptoacetic acid and ethylene-dimercaptan is possible without the usual accompanying oxidative decomposition.

ii) Six base equivalents per mol of intermediate which are replaceable and volatile in acid allow the use of acidic ligands such as amine hydrochlorides, carboxylic acids, phenols and mercaptans.

iii) The intermediate is stable on storage if kept dry and may be prepared in quantity. If sufficient acid is present, the evolution of CO_2 gas completes the reaction even under relatively mild conditions.

This general method was then extended not only to complexes with other bidentate ligands but also to those with a terdentate or a sexadentate ligand, e.g. bidentates, (+)-hydroxymethylenecamphor [46], (+)-3-acetylcamphor [47] and (−)-cyclohexanediamine [48]; terdentates, 2,3-diaminopropionic acid [49], 2,4-diamino-butyric acid [50], 1,3-diamino-2-propanol [51], and 1-methyl-2,4,6-triaminocyclohexane [52]; and sexadentates, 5-ethyl-5-(δ-amino-β-azabutyl)-1,9-diamino-3,7-diazanonate [53], ethylenediamino-*N,N'*-disuccinic acid [54] and (S)-ethylenediamine-*N,N'*-diacetic-*N'*-monosuccinic acid [55].

Tris-(+)-hydroxymethylenecamphoratocobalt(III) is soluble in common organic solvents [46]; stoichiometric amounts of the ligand and freshly prepared $Na_3[Co(CO_3)_3]$ · 3 H_2O in a 50:50 mixture of water and benzene are shaken at room temperature overnight. The tris-complex is separated from the benzene-insoluble carbonato complex. The green solid, obtained by evaporating the benzene layer, is treated several times with ligroin (b.p. 40～60 °C) to separate the ligroin-soluble tris-complex from the ligroin-insoluble bis(hydroxymethylenecamphor)cobalt(II) complex. The pure tris-complex is recrystallized from the ligroin solution. This procedure can also be used to prepare the tris-complex of (+)-3-acetylcamphorate [47].

2.5 General Syntheses of Amino Acid Complexes

2.5.1 Tris-Complexes with an Amino Acid

When three molecules of an α-amino acid, $NH_2CH(R)$ COOH, coordinate to cobalt(III) through the α-amino nitrogen and α-carboxylate oxygen, two geometrical isomers are possible and they are classified as *fac* (or *cis-cis*) and *mer* (or *trans-cis*) according to the arrangement of the three identical donor atoms. Three preparative methods [56] gave different yields of the geometrical isomers of each tris-complex of glycine, racemic alanine and L-leucine. The results were as follows:

i) The classic method of Ley and Winkler [57], dissolving freshly prepared cobalt(III) hydroxide oxide, CoO(OH), in an aqueous solution of an amino acid under heating, gives products with the *mer* isomer predominating.

ii) The method originated by Neville and Gorin [58], allowing an amino acid to react with hexaamminecobalt(III) chloride in a boiling aqueous KOH solution under reflux, favors formation of the *fac* isomer.

iii) The method devised by Shibata et al. [56)], allowing an amino acid to react with "the green solution" of potassium tricarbonatocobaltate(III) gives both isomers in approximately equal amounts.[7]

Several years later, the preparative work [56)] on the geometrical isomers was renewed by other workers, who applied the three methods to the preparation of geometrical-optical isomers, or diastereoisomers, of tris-complexes from optically active amino acids. In general, four diastereoisomers are possible for a tris(L- or D-amino-acidato) complex: *fac*(+), *fac*(−), *mer*(+) and *mer*(−).

Dealing with optical rotatory properties of the isomers of tris(L-alaninato)-cobalt(III), Dunlop and Gillard [59)] reported that cobalt(III) oxide hydroxide gives predominantly two *mer*-diastereoisomers, hexaamminecobalt(III) chloride gives only water-insoluble *fac*(+)-isomer, and starting from tricarbonatocobaltate(III) complex results in the isolation of all isomers, *fac*(+), *mer*(+), *fac*(−) and *mer*(−), the last two by chromatography on alumina. The three isomers [60)] from the Ley-Winkler method were elucidated to be in fact pure *fac*(+), *mer*(+) and impure *mer*(−) [59)]. The four isomers were isolated from tricarbonatocobaltate(III) [61)] as by the Ley-Winkler method [62)].

Denning and Piper [63)] studied optical activity, absolute configuration, and rearrangement reactions of tris-complexes with L-alanine, L-leucine, and L-proline. They modified the method of Shibata et al. by the use of an amino acid dissolved in aqueous hydrochloric acid.[8] By this method, *mer*(+) and *mer*(−) isomers of [Co(L-leu)₃] were effectively obtained, while *fac*(+) and *fac*(−) were obtained by the method of Neville. Three of four isomers of [Co(L-pro)₃], *fac*(+), *fac*(−) and *mer*(−) obtained by the above-mentioned modified method were chromatographed on alumina. No material was identifiable as the *mer*(+)-isomer; this is attributed to the prohibitive steric hindrance due to two L-proline residuals (see 4.2). The method using hexaamminecobalt(III) salt failed to give any product after prolonged refluxing. The reaction between tetraamminediaquacobalt(III) complex and L-proline with activated charcoal [64)] gives rise to only one isomer, *fac*(−).

The geometrical isomers of the tris-complex with β-alanine were synthesized by Ćelap et al. [65)], since the method of Ley-Winkler failed. With the hexaamminecobalt-(III) complex the product was *fac*-isomer with a small amount of *mer*-isomer. The reaction between sodium tricarbonatocobaltate(III) trihydrate and β-alanine in aqueous solution under heating yielded *mer*-isomer effectively.

---

[7] Method using potassium tricarbonatocobaltate(III): An amino acid is poured into "the green solution", and the mixture is warmed on a water-bath until the color of the solution changes from green to blue. Then 6 mol/dm³ acetic acid is added dropwise until the evolution of carbon dioxide ceases and the solution becomes red-violet. After filtering, the solution stands for some time in the cold. Crystals of the less soluble *fac*-isomer are deposited, then the *mer*-isomer separates out on evaporating the mother liquor over sulfuric acid.

[8] Preparation of *mer*-tris(L-leucinato(cobalt(III) isomers: $Co(NO_3)_2 \cdot 6\,H_2O$ (17.4 g, 0.06 mol) in 30 cm³ water and 6 cm³ 30% H_2O_2 are mixed and added dropwise ro 30.3 g of $KHCO_3$ (0.3 mol) in 30 cm³ of H_2O at 0 °C. After stirring the mixture at 0 °C for 1 hr, 23.7 g of L-leucine (0.18 mol) dissolved in 180 cm³ of 1 mol/dm³ HCl (0.18 mol) is added slowly while stirring and the mixture is heated cautiously to boiling and refluxed for 2 hrs. After cooling the crude product is filtered and dried at 100 °C; yield (crude) 24.2 g (89.5%). The crude material is extracted three times by boiling ethanol and the filtrate is evaporated to dryness. The product is dissolved in minimum 85% ethanol-water and chromatographed on an alumina column (140 × 4 cm) (see 3.4.2).

Two *mer*-isomers of the tris-complex with L-aspartic acid [66] were isolated by the method of Ley-Winkler. Shibata et al. [67], however, isolated all four isomers as the hydrogen L-aspartato complex, [Co(L-Hasp)₃], by allowing L-aspartic acid to react with "the green solution" in the presence of activated charcoal at room temperature, and then by lowering the pH values with aqueous hydrochloric acid. The *fac*(−), *mer*(−), *fac*(+) and *mer*(+) precipitated in turn. Warming "the green solution" with L-aspartate in the absence of activated charcoal results in the formation of bis-complex containing aspartate anions as terdentate ligand [68].

Thus, within only a few years in the 1960's, the syntheses of geometrical and optical isomers of the tris-complexes with amino acids rapidly advanced with absorption, circular dichroism, and proton magnetic spectral studies as well X-ray studies. The synthesis using tricarbonatocobaltate(III) proved to be very useful for tris-complexes of amino acids.

2.5.2 Tris-Complexes with two Kinds of α-Amino Acids

Shibata et al. [69] proposed a general procedure for the synthesis of mixed amino-acidato complexes using "the green solution" of tricarbonatocobaltate(III). The procedure was based on the fact that "it is more difficult for a ligand to be substituted for the third carbonate ligand of $[Co(CO_3)_3]^{3-}$ than for the first and second carbonate ligands." That is, in the first step a bis(amino-acidato)carbonatocobaltate(III) complex is produced and then the reaction with another amino acid leads to the formation of the desired mixed amino-acidato complex, e.g.

$$K[Co(CO_3) (gly)_2] \cdot H_2O,^9 \qquad K[Co(CO_3) (\text{L-ala})_2] \cdot 3 H_2O,$$

$$[Co(gly)_2 (\text{L-val})],^{10} \qquad fac(+)\text{-}[Co(fly) (\text{L-val})_2],$$

$$fac(+)\text{-}[Co(\text{L-ala}) (\text{L-val})_2], \qquad fac(+)\text{-}[Co(gly) (\text{L-val})_2],$$

$$fac(+)\text{-}[Co(\text{L-ala}) (\text{L-val})_2], \qquad fac(-)\text{-}[Co(\text{D-ala}) (\text{L-val})_2],$$

---

[9] Preparation of potassium carbonatobis(glycinato)cobaltate(III): "The green solution" prepared from 0.1 mol CoCl₂ · 6 H₂O with added glycine (15 g, 0.2 mol) and water (60 cm³), is stirred at 60 °C for 3 hrs. The resulting violet solution is filtered, cooled to room temperature and then adjusted tp pH 7.2 ~ 7.5 with aqueous acetic acid. After adding some ethanol, the mixture is kept cold overnight in order to precipitate potassium chloride and potassium bicarbonate. After filtration, ethanol (*ca.* 200 cm³) is added to the filtrate in the cold. When scratching red-violet crystals gradually deposit. The crystals are collected and recrystallized from water. The yield is about 12 g (40%).

When a supersaturated aqueous solution of this compound was left to stand at room temperature, spontaneous resolution took place [70].

[10] Preparation of bis(glycinato)L-valinatocobalt(III): The carbonatobis(glycinato) complex (6.4 g, 0.02 mol) is dissolved in water (100 cm³); L-valine (2.3 g, 0.02 mol) is added to the solution and stirred at 60 °C for several hours. By cooling, the β(+)-isomer, insoluble in water, is obtained. When the filtrate is mixed with ethanol (*ca.* 50 cm³), the second isomer, β(−), separates out and is recrystallized from water as red needles. The remaining ethanolic solution is concentrated until the third isomer, *mer*(+), deposits as red-violet plates. They are recrystallized from water. The final liquor from the above is evaporated to dryness and the residue is extracted with methanol for chromatography on alumina; a violet band is observed, but the fourth *mer*(−)isomer is not obtained as crystals because of its high hygroscopicity. The yields are 0.3 g (*fac*(+), anhydrate), 0.5 g (*fac*(−), 1.5-hydrate), and 1 g (*mer*(+), 2-hydrate).

and

$$fac(+)\text{-}[Co(\text{D-ala})\,(\text{L-ala})_2]$$

Kostić and Niketić [71)] used this method to prepare geometrical isomers of the mixed cobalt(III) complexes containing glycinate and β-alaninate ions, $[Co(\beta\text{-ala})_x(gly)_{3-x}]$ ($x = 1, 2$); the reaction of β-alanine or glycine, in the presence of activated charcoal, with $[Co(CO_3)(gly)_2]^-$ or $[Co(CO_3)(\beta\text{-ala})_2]^-$ is more adequate for isomers than the reaction without activated charcoal.

Attempts to synthesize a complex with three kinds of amino acids have been made in our laboratory. An intermediate $[Co(CO_3)(\text{L-aspn})(\text{L-lys}^+)]$ (aspn represents the asparaginate dianion and lys$^+$, protonated lysine) was prepared by allowing L-asparagine and L-lysine to react with "the green solution". After acid hydrolysis, the resulting diaqua complex was allowed to react with glycine to give the aimed complex $[Co(\text{L-aspn})(gly)(\text{L-lys}^+)]^+$, which was chromatographed on an ion-exchange column to separate the distereoisomers [72)].

2.6 Binuclear Complexes

The polynuclear complexes of cobalt(III), systematized by Werner, have now found new interests in reaction mechanisms, bonding and structures, preparation and so on. With respect to binuclear cobalt(III) complexes, a comprehensive review covering the literature up to 1970 was published by Sykes and Weil [73)]. Procedures for the preparation of some fundamental dicobalt complexes are found in Inorganic Syntheses [74)].

The most common bridging ligands are O_2^{2-}, O_2^-, NH_2^- and OH^-. In general, the μ-peroxo complex can be prepared by air-oxidation of an ammonical solution of a cobalt(II) salt, the μ-superoxo complex by oxidation of the corresponding peroxo complex with an oxidizing agent, the μ-amido-μ-peroxo complex by condensation reaction of the corresponding μ-peroxo complex involving ammine ligands, and the μ-hydroxo complex by condensation reaction of a mononuclear complex involving aqua ligands.

2.6.1 μ-Carbonato and μ-Oxalato Complexes

Well-established examples of binuclear cobalt(III) complexes with carbonate bridges are very limited. Kremer and Mac-Coll [75)] confirmed the existence of the μ-carbonato-bis(pentaamminecobalt(III)) complex, $[(NH_3)_5\,Co(CO_3)\,Co(NH_3)_5]^{4+}$ (originally prepared by Kranig [76)]) by potentiometric titration, conductometric charge determination, visible, ultraviolet, and infrared spectra, behavior against ion exchange resin, counter-ion substitution, thermogravimetric analysis and differential thermal analysis.

Nishide and Saito [77)], through the reaction of (S)-2-amino-1-propanol (S-propanol amine, S-Hpra) with "the green solution" of tricarbonatocobaltate(III), prepared a few binuclear complexes with carbonate bridges, from which a few binuclear complexes with oxalate bridges were derived. The preparation is shown in Scheme I,

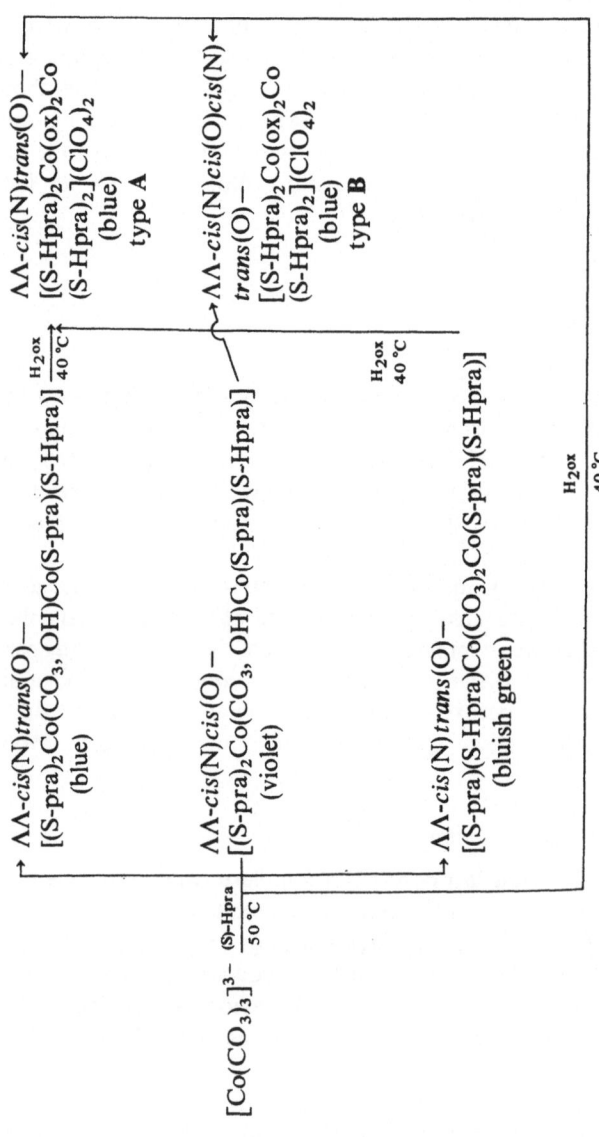

Scheme 1. Outline of the Preparation

where type **A** indicates coordination of two Co(III) through two O atoms on one C atom, type **B**, through two O atoms on different carbon atoms; these are distinguished mainly by IR spectra.

2.6.2 μ-Hydroxo Complexes

The binuclear complexes of the formula $[(am)_2M(\mu\text{-}OH)_2M(am)_2]$ (am represents an amino-acidate ion) were rather familiar for M = Cr(III) [78,79,80]. The complexes of Co(III) with am = gly, L-ala, or L-val were reported by Gillard et al. [81]. They used a bis(aminoacidato)carbonatocobaltate(III) complex, $[Co(CO_3)(am)_2]^-$, prepared from "the green solution" and an amino acid, as the starting material. The conversion of a carbonate ion into an aqua ligand by acidification with aqueous perchloric acid gave the dimeric complex.

Twenty-one geometrical-optical isomers are possible for a complex of this type. However, the actually isolated compound was only one for each amino acid. The compounds are: tetrakis(glycinato)-di-μ-hydroxodicobalt(III, III) monohydrate, (+), (+)-tetrakis(L-alaninato)-di-μ-hydroxodicobalt(III, III), and (+),(+)-tetrakis(L-vali-nato)-di-μ-hydroxodicobalt(III, III).

2.7 Heteropoly Electrolytes

Two kinds of compounds containing cobalt(III) and molybdenum(VI) were prepared [82], one by oxidizing a mixed solution of cobalt(II) acetate and ammonium para-molybdate with ammonium persulfate, and another by oxidizing a similar solution with hydrogen peroxide. The former was regarded as a dualistic "oxide" $3\,(NH_4)_2O \cdot Co_2O_3 \cdot 12\,MoO_3 \cdot 20\,H_2O$ and the latter as $2\,(NH_4)_2O \cdot Co_2O_3 \cdot 10\,MoO_3 \cdot 12\,H_2O$.

Baker et al. [83] prepared the above-mentioned 1:6 compound by adding an aqueous solution of tetraamminecarbonatocobalt(III) nitrate to a boiling solution of ammonium paramolybdate. They formulated the compound as $(NH_4)_2H_6$-$[Co(MoO_4)_6] \cdot 7\,H_2O$ according to Miolati-Rosenheim. The next proposed formula [84] $[(XO_6Mo_6O_{15})_n]^{3n-}$, where X represents Cr(III), Fe(III), Co(III) or Al(III) and n is an undetermined integer which is probably small, was based on potentiometric titrations, dehydration experiments, magnetic measurements etc. Furthermore, the correct formula [85] was claimed for the ammonium salt of the chromium(III) complex to be $(NH_4)_6[(CrO_6Mo_6O_{15})_2] \cdot 20\,H_2O$; all the compounds of Cr(III), Fe(III), Co(III) and Al(III) are isomorphous on the X-ray diffraction photographs.

Absorption studies with those compounds [86] revealed that the spectra in the visible region were very similar to those of the corresponding hexaaqua complexes

$[M(H_2O)_6]^{3+}$ (M = Al, Cr, Fe, Co), so the existence of a chromophore $[M(O)_6]$ in every heteropoly anion was claimed. The absorption spectrum was measured for $2\,(NH_4)_2O \cdot Co_2O_3 \cdot 10\,MoO_3 \cdot 12\,H_2O$, too; since the spectrum exhibited an absorption band characteristic of a μ-hydroxo polynuclear complex in addition to d-d absorption bands, the compound was considered to have a kind of polynuclear structure.

Baker and McCutcheon [87] discovered four kinds of heteropoly salts containing cobalt and tungsten in the anions; the addition of aqueous cobalt(II) acetate to a boiling solution of sodium tungstate, adjusted to pH 6.5 ∼ 7.5 with acetic acid, and then the addition of a hot saturated solution of potassium chloride as precipitant gave dark emerald green crystals formulated as $K_8[Co^{2+}Co^{2+}W_{12}O_{42}] \cdot 15\,H_2O$. Adding potassium persulfate to the mentioned reaction mixture and then potassium nitrate as precipitant gave dark brown crystals with the formula $K_7[Co^{2+}Co^{3+}W_{12}O_{42}] \cdot 16\,H_2O$. The octa potassium salt was dissolved in dil. hydrochloric acid; after evaporation green crystals of $K_5H_5[Co^{2+}(W_2O_7)_6] \cdot 16\,H_2O$ were obtained. When the octa potassium salt was dissolved in dil. sulfuric acid boiled, and oxidized with potassium persulfate light yellow crystals of $K_4H_5[Co^{3+}(W_2O_7)_6] \cdot 18\,H_2O$ separated out.

Shimura and Tsuchida [88] measured the absorption; the spectrum of $[Co^{2+}(W_2O_7)_6]^{10-}$ was characteristic of tetrahedrally coordinated cobalt(II), and hence the formula $[Co^{II}O_4W_{12}O_{38}]^{10-}$ or $[Co^{II}O_4W_{12}O_{36}]^{6-}$ was assumed. The spectrum of $[Co^{III}(W_2O_7)_6]^{9-}$ was entirely different from the spectra of hexa-coordinated Co(III) complexes, a broad maximum being observed at $25\,660\ cm^{-1}$ with intensity $\log \varepsilon = 3.10$. From this result, $[Co^{III}O_4W_{12}O_{38}]^{9-}$ or $[Co^{III}O_4W_{12}O_{36}]^{5-}$ was tentatively assigned. The spectrum of $[Co^{II}Co^{II}W_{12}O_{42}]^{8-}$ revealed that it contains at least one cobalt(II) atom in tetrahedrally coordinated state, therefore, the tentative formula $[Co^{II}O_4(W_{12}O_{32})\,O_6Co^{II}]^{8-}$ was assigned. The spectrum of $[Co^{II}Co^{III}W_{12}O_{42}]^{7-}$ exhibited a complicated structure, a broad band at ca. $26\,000\ cm^{-1}$ ($\log \varepsilon = 3.20$) being an evidence for tetrahedrally coordinated cobalt(III). Thus, the formula of this anion was regarded as $[Co^{III}O_4(W_{12}O_{32})\,O_6Co^{II}]^{7-}$.

Baker et al. [89] established a new general structural category of heteropoly anions, formulated as

$$[H_hM^{m+}O_6X^{x+}O_4W_{11}O_{30}]^{(14-m-x-h)-}$$

The structure is a modification of the well-known 12-tungsto "Keggin" structure. Octahedrally coordinated M^{m+} replaces just one of the 12 octahedral W atoms of the conventional Keggin structure, and X^{x+} occupies the Keggin unit's central tetrahedral cavity. In some cases X consists of two H atoms, fulfilling presumably the same role as those in the metatungstate ion, $[H_2W_{12}O_{40}]^{6-}$, which has the Keggin structure. The number of other H atoms, H_h (which must be firmly attached to exterior oxygen atoms of the complex, most probably to those surrounding M), seems to be mainly characteristic of the identity of M^{m+}.

Fifteen salts were classified into the following five series;

anion 1: M = Co^{2+} , X = Si^{4+}

anion 2: M = Co^{3+} , X = H_2^{2+}

anion 3: $M = Ga^{3+}$, $X = H_2^{2+}$

anion 4: $M = Co^{2+}$, $X = Co^{2+}$

anion 5: $M = Co^{2+}$, $X = Co^{3+}$

where anions 4 and 5 originally erroneously formulated as 12-tungstodicobaltates [87].

The compound with anion 2 was prepared by stirring "the green solution" into an ice-cooled solution of sodium tungstate whose pH had been adjusted to 5 with acetic acid. [90] The continued stirring produced yellow-green crystals formulated as $K_7[CoO_6W_{11}O_{38}] \cdot 17 H_2O$.[11] The preparation of this 11-tungstocobaltate(III) led to the discovery of the other 11-tungsto compounds.

"The green solution" produced another new compound formulated as $K_5[CoW_6-O_{22}] \cdot 7 H_2O$ in a similar procedure. The known compounds 6-molybdocobaltate(III) and 10-molybdodicobaltate(III) were also easily prepared from "the green solution" with aqueous ammonium paramolybdate [31]. Evans and Showell [91] determined crystal structure of this decamolybdodicobaltate(III) salt and its proper formula $(NH_4)_6-[H_4Co_2Mo_{10}O_{38}] \cdot 7 H_2O$.

2.8 Reaction Kinetics Concerning Tricarbonatocobaltate(III)

The versatility of the Field-Durrant solution for the synthesis of cobalt(III) complexes depends largely upon high lability of tricarbonatocobaltate(III) in the substitutions, but kinetic studies concerning such labile nature are limited at present. However, pioneering papers by Davies and Hung studied the substitutions of tricarbonato-cobaltate(III) with pyridine [92], ethylenediamine [93], 1,2-propanediamine (1,2-pn) [93], and 1,3-propanediamine (1,3-pn) [93]. A solution prepared by dissolving $Na_3[Co(CO_3)_3] \cdot 3 H_2O$ in aqueous $NaHCO_3$ is sufficiently stable to allow accurate kinetic measurements to be made and is suitable for kinetic studies.[12]

"The green solution" reacts with excess pyridine only to produce cis-$[Co(CO_3)_2-(py)_2]^-$ under the experimental conditions; $[Co^{III}] = 5.7 \sim 11.4) \times 10^{-4}$ mol/dm^3,

---

[11] Potassium 11-tungstocobaltate(III): Sodium tungstate, $Na_2WO_4 \cdot 2 H_2O$, 100 g is dissolved in 160 cm^3 of hot water and the solution is adjusted to pH $ca.$ 5 with glacial acetic acid (about 34 cm^3) in an ice-bath. "The green solution" prepared from 6 g $CoCl_2 \cdot 6 H_2O$ is slowly added to the cooled solution with constant stirring. During further stirring a small amount of white precipitate separates out, which is filtered off. After 2 hrs, a considerable amount of the crude product is obtained as a yellow-green precipitate. The precipitate is dissolved in minimum warm water; then the solution is kept in a refrigerator after adding 5 g potassium chloride. The dissolution in water and the addition of potassium chloride are repeated several times. Finally yellow-green cubic crystals are obtained in pure state. Final yield less than 2 g.

[12] Preparation of the stock solution [92]: $Na_3[Co(CO_3)_3] \cdot 3 H_2O$ is synthesized by the method of Bauer and Drinkard [12] (see 2 · 2) and stored over P_2O_5 in the dark, since moisture causes decomposition to a black, insoluble residue. A saturated stock solution is prepared by adding a small amount of the solid to 1.0 mol/dm^3 $NaHCO_3$ and stirring the mixture for $ca.$ 40 min. After filtration through a glass filter and staying overnight at room temperature, solution is refiltered. It contains $[Co^{III}] 3 \times 10^{-3}$ mol/dm^3 and is stored at 0 °C. Spectral maxima at 260 nm ($\varepsilon = 10^5$ M^{-1} cm^{-1}), 440 nm ($\varepsilon = 166 \pm 2$ M^{-1} cm^{-1}), and 635 nm ($\varepsilon = 154 \pm 2$ M^{-1} cm^{-1}); Beer's law is obeyed in the range of $8 \lesssim pH \lesssim 10$.

$[H^+] = (0.15 \sim 9.52) \times 10^{-9}$ mol/dm^3, $[HCO_3^-] = 0.14 \sim 0.865$ mol/dm^3, $[py] = = (1.2 \sim 29.8) \times 10^{-2}$ mol/dm^3 at 25 °C, and ionic strength 1.0 mol/dm^3 (NaHCO$_3$, NaClO$_4$). The rate law was found to be

$$- d[Co(CO_3)_3^{3-}]/dt = d[cis\text{-}Co(CO_3)_2(py)_2^{2-}]/dt$$

$$= \left(\frac{A + B/[HCO_3^-][H^+]}{1 + C[H^+]}\right) [Co(CO_3)_3^{3-}][py] \tag{1}$$

where A, B and C were empirical parameters. The dependence of the reaction rate on acidity and free bicarbonate concentration was interpretated by the following equations.

$$Co(CO_3)_3^{3-}(aq) + H_3O^+ \overset{fast}{\rightleftharpoons} Co(CO_3)_2(HCO_3)(H_2O)^{2-} \quad K_h \tag{2}$$

$$Co(CO_3)_2(HCO_3)(H_2O)^{2-} \overset{fast}{\rightleftharpoons} Co(CO_3)_2(H_2O)_2^- + HCO_3^- \quad K_{H_2O} \tag{3}$$

The equilibrium constants K_h and K_{H_2O} were found to be $(1.31 \pm 0.12) \times 10^9$ dm^3/mol and 0.06 mol/dm^3 at 25 °C, respectively.

Thus, the incoming py ligand was regarded to substitute for a coordinated water molecule of the aqua complex produced in the reactions (2) and (3), the fully chelated species, $[Co(CO_3)_3]^{3-}$, being much more inert to substitution. In the following equations,

$$Co(CO_3)_3^{3-} + py \xrightarrow{k_0} Co(CO_3)_2(CO_3)(py)^{3-} \tag{4}$$

$$Co(CO_3)_2(HCO_3)(H_2O)^{2-} + py \xrightarrow{k_1} Co(CO_3)_2(HCO_3)(py)^{2-} + H_2O \tag{5}$$

$$Co(CO_3)_2(H_2O)_2^- + py \xrightarrow{k_2} Co(CO_3)_2(H_2O)(py)^- + H_2O \tag{6}$$

$$Co^{III}py + py \xrightarrow[fast]{k_3} Co(CO_3)_2(py)_2^- \tag{7}$$

the rate constants were estimated to be $k_0 \leq 0.4$, $k_1 = 1.3 \pm 0.2$ and $k_2 \geq 10$ dm^3/mol min.$^{-1}$, respectively. The last equation represents the rapid reactions of the three initial carbonatopyridinecobaltate(III) products with pyridine to give cis-[Co(CO$_3$)$_2$-(py)$_2$]$^-$.

In preliminary studies using the green stock solution in 1 mol/dm^3 NaHCO$_3$, a substantial, immediate, and reproducible absorption increase in the near-UV region occurred on addition of excess en, 1,2-pn, and 1,3-pn dissolved in aqueous NaHCO$_3$. These brown solutions slowly became purple, exhibiting visible absorption spectra characteristic of [Co(CO$_3$)$_2$(diamine)]$^-$ species. By contrast, addition of ammonia, 1,6-hexanediamine, and pyridine under similar conditions did not result in such brown coloration; the color of the solutions simply changed from green to purple over periods of 5~60 min., inspiring the workers to study the kinetics of the reactions between [Co(CO$_3$)$_3$]$^{3-}$ and en, 1,2-pn, and 1,3-pn in aqueous NaHCO$_3$.

The kinetic processes observed on mixing the green cobalt(III) reactant with the diamines were monitored by stopped-flow and conventional spectrophotometry.

The conditions employed were: $[Co^{III}] = (2.31 \sim 3.99) \times 10^{-4}$ mol/dm³, $[H^+] = (0.06 \sim 2.81) \times 10^{-9}$ mol/dm³, $[HCO_3^-] = 0.02 \sim 0.68$ mol/dm³, $[en]_T = (2.3 \sim 95.0) \times 10^{-3}$ mol/dm³, $[1,2\text{-pn}]_T = (4.68 \sim 110.0) \times 10^{-3}$ mol/dm³, and $[1,3\text{-pn}]_T = (2.6 \sim 150.0) \times 10^{-3}$ mol/dm³, at $8 \leq$ pH ≤ 10.5 and 25 °C, and ionic strength 1.0 mol/dm³.

From the spectral changes after mixing the reactants, the following scheme was proposed;

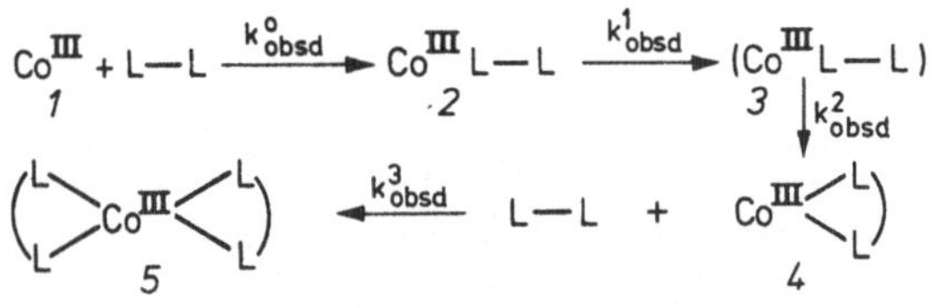

Here *1* is the initial, green carbonatocobaltate(III) reactant, L-L is the bidentate diamine ligand, *2* and *3* are the initial brown product and brown derivative thereof, *4* is the purple $[Co(CO_3)_2(diamine)]^-$ chelate, and *5* is the red $[Co(CO_3)(diamine)_2]^+$ chelate, respectively. The rate constants k^1_{obsd}, k^2_{obsd}, and k^3_{obsd} were the computer-calculated first-order rate constants for the respective steps. The rates of the reactions $1 \to 2$ (called half-bonded complex formation), $2 \to 3$ (relaxation), and $3 \to 4 \to 5$ (ring closure) decreased progressively with a given ligand. The half-bonded diamine products, $Co(CO_3)_2(HCO_3)(diamine)^{2-}$(aq), exhibited unusually large molar absorptivities ($\varepsilon \approx 1100$ dm³/mol cm⁻¹ at 380 nm), which were not apparent in the corresponding reactions with NH_3, py or 1,6-hexanediamine. From the kinetic data for the rapid formation of these half-bonded diamine intermediates an associative interchange mechanism involving unusually stable reaction precursors with en and 1,2-pn was proposed.

Tanner [94] investigated the kinetics of the autocatalysis by Co(II) in the reduction of carbonatocobaltate(III) complexes by hydrazine, in which saturated solutions of sodium tri(carbonato)cobaltate(III) trihydrate in 1 mol/dm³ $NaHCO_3$ were used for the stock solution.

2.9 References

1. Field, F.: J. Chem. Soc. *14*, 48 (1862)
2. Durrant, R. G.: J. Chem. Soc. Proc. *12*, 96, 244 (1896)
3. Durrant, R. G.: J. Chem. Soc. *87*, 1781 (1905)
4. Job, A.: Compt. Rend. *127*, 100 (1898)
5. Job, A.: Ann. Chim. Phys. *20*, 205 (1900)
6. Duval, C.: Compt. Rend. *191*, 615 (1930)
7. Duval, C.: Anal. Chim. Acta *1*, 201 (1947)
8. Laitinen, H. A., Burdett, L. W.: Anal. Chem. *23*, 1268 (1951)
9. McCutcheon, T. P., Schule, W. J.: J. Am. Chem. Soc. *75*, 1845 (1953)
10. Mori, M., Shibata, M.: J. Chem. Soc. Jap. *75*, 1046 (1954)
11. Mori, M., Shibata, M., Kyuno, E., Adachi, T.: Bull. Chem. Soc. Jpn. *29*, 883 (1956)
12. Bauer, H. F., Drinkard, W. C.: J. Am. Chem. Soc. *82*, 5031 (1960)
13. Mac-Coll, C. R. P.: Coord. Chem. Rev. *4*, 147 (1969)
14. Krishnamurty, K. V., Harris, G. M., Sastri, V. S.: Chem. Rev. *70*, 171 (1970)

15. Shibata, M.: Proc. Jap. Acad. *50*, 779 (1974)
16. Inorg. Synth. *8*, 202 (1966)
17. Inorg. Synth. *10*, 45 (1967)
18. Baur, J. A., Bricker, C. E.: Anal. Chem. *37*, 1461 (1965)
19. Jørgensen, C. K.: Absorption spectra and chemical bonding in complexes. London: Pergamon 1962
20. Telep, G., Boltz, D. F.: Anal. Chem. *24*, 945 (1952)
21. Duval, R., Duval, C., Lecompte, J.: Bull. Soc. Chim. France *10*, 517 (1943)
22. Gatehouse, B. M., Livingstone, S. E., Nyholm, R. S.: J. Chem. Soc. *1958*, 3137
23. Fujita, J., Martell, A. E., Nakamoto, K.: J. Chem. Phys. *36*, 339 (1962)
24. Nakamoto, K.: Infrared spectra of inorganic and coordination compounds. 2nd ed. Wiley-Interscience 1970
25. Lascombe, J.: J. Chim. Phys. *56*, 79 (1959)
26. Nakamoto, K., Fujita, J., Tanaka, S., Kobayashi, J.: J. Am. Chem. Soc. *79*, 4906 (1957)
27. Gillard, R. D., Mitchel, P. R., Price, M. G.: J.C.S. Dalton *1972*, 1211
28. Gillard, R. D., Shepherd, D. J., Tarr, D. A.: J.C.S. Dalton *1976*, 594
29. Segupta, A. K., Nandi, A. K.: Z. Anorg. Allg. Chem. *404*, 81 (1974)
30. Segupta, A. K., Nandi, A. K.: Z. Anorg. Allg. Chem. *403*, 327 (1974)
31. Shibata, M.: Nippon Kagaku Zasshi *87*, 771 (1966)
32. Mori, M., Shibata, M., Kyuno, E., Hoshiyama, K.: Bull. Chem. Soc. Jpn. *31*, 291 (1958)
33. Rowan, N. S., Storm, C. B., Hunt, J. B.: Inorg. Chem. *17*, 2853 (1978)
34. Ida, Y., Imai, K., Shibata, M.: Bull. Chem. Soc. Jpn. *51*, 2741 (1978)
35. Kashiwabara, K., Igi, K., Douglas, B. E.: Bull. Chem. Soc. Jpn. *49*, 1573 (1976)
36. Schramm, W.: Z. Anorg. Chem. *180*, 167, 177 (1929)
37. Riesenfeld, E. H., Klement, R.: Z. Anorg. Chem. *124*, 14 (1922)
38. Dwyer, F. P., Reid, I. K., Garvan, F. L.: J. Am. Chem. Soc. *83*, 1285 (1961)
39. Ichikawa, H., Shibata, M.: Bull. Chem. Soc. Jpn. *42*, 2873 (1969)
40. Ichikawa, H., Shibata, M.: Bull. Chem. Soc. Jpn. *43*, 3789 (1970)
41. Golovnya, V. A., Kokl, L. A., Sokol, C. K.: Russ. J. Inorg. Chem. *10*, 829 (1965)
42. Golovnya, V. A., Kokl, L. A., Sokol, C. K.: Russ. J. Inorg. Chem. *10*, 836 (1965)
43. Fujinami, S., Shibata, M.: Chem. Lett. *1972*, 219
44. Fujinami, S., Shibata, M.: Bull. Chem. Soc. Jpn. *43*, 3789 (1970)
45. Hyodo, O., Archer, R. D.: Inorg. Chem. *8*, 2510 (1969)
46. Dunlop, J. H., Gillard, R. D.: J. Chem. Soc. *1965*, 6531
47. King, R. M., Everett, G. W., Jr.: Inorg. Chem. *10*, 1237 (1971)
48. Sarneski, J. E., Urbach, F. U.: J. Am. Chem. Soc. *93*, 884 (1971)
49. Freeman, W. A., Liu, C. F.: Inorg. Chem. *7*, 764 (1968)
50. Freeman, W. A., Liu, C. F.: Inorg. Chem. *9*, 1191 (1970)
51. Okamoto, M. S., Barefield, E. K.: Inorg. Chem. *13*, 2611 (1974)
52. Reeman, W. A., Liu, C. F.: Inorg. Chem. *14*, 2120 (1975)
53. Green, R. W., Catchpole, K. W., Phillip, A. T., Lions, F.: Inorg. Chem. *2*, 597 (1963)
54. Neal, J. A., Rose, N. J.: Inorg. Chem. *7*, 2405 (1968)
55. Legg, J. I., Neal, J. A.: Inorg. Chem. *12*, 1805 (1973)
56. Mori, M., Shibata, M., Kyuno, E., Kanaya, M.: Bull. Chem. Soc. Jpn. *34*, 1837 (1961)
57. Ley, H., Winkler, H.: Ber. *42*, 3894 (1909)
58. Neville, R., Gorin, G.: J. Am. Chem. Soc. *78*, 4893 (1956)
59. Dunlop, J. H., Gillard, R. D.: J. Chem. Soc. (A) *1965*, 6531
60. Lifschitz, I.: Proc. k. ned. Akad. Wetenschap. *15*, 721 (1924)
61. Larsen, E., Mason, S. F.: J. Chem. Soc. (A) *1966*, 313
62. Douglas, B. E., Yamada, S.: Inorg. Chem. *4*, 1651 (1965)
63. Denning, R. G., Piper, T. S.: Inorg. Chem. *5*, 1056 (1966)
64. Yasui, T., Hidaka, J., Shimura, Y.: Bull. Chem. Soc. Jpn. *38*, 2025 (1965)
65. Ćelap, M. B., Niketić, T. J., Nibolić, T. J., Nibolić, V. N.: Inorg. Chem. *6*, 2063 (1967)
66. Lifschitz, I., Froentjes, W.: Rec. Trav. Chim. *60*, 225 (1941)
67. Shibata, M., Nishikawa, H., Hosaka, K.: Bull. Chem. Soc. Jpn. *40*, 236 (1967)
68. Hosaka, K., Nishikawa, H., Shibata, M.: Bull. Chem. Soc. Jpn. *42*, 277 (1969)
69. Shibata, M., Nishikawa, H., Nishida, Y.: Inorg. Chem. *7*, 9 (1968)
70. Kawasaki, K., Yoshii, J., Shibata, M.: Bull. Chem. Soc. Jpn. *43*, 3819 (1970)

71. Kostić, N. M., Niketić, S. R.: J.C.S. Chem. Commu. *1977*, 676
72. Fujinami, S., Shibata, M.: unpublished
73. Edwards, J. O. (ed.): Inorganic reaction mechanisms. Progress in inorganic chemistry vol. 13. Interscience Pub. 1970, pp. 2–106
74. Davies, R., Mori, M., Sybes, A. G., Weil, J. A.: Inorganic Syntheses *12*
75. Kremer, E., Mac-Coll, C. R. P.: Inorg. Chem. *10*, 2182 (1971)
76. Kranig, J.: Ann. Chim. Paris *41*, 87 (1929)
77. Nishide, T., Saito, K.: Bull. Chem. Soc. Jpn. *50*, 2618 (1977)
78. Ley, H., Ficken, K.: Ber. *45*, 377 (1912)
79. Earnshaw, A., Lewis, J.: J. Chem. Soc. *1961*, 396
80. Husain, M., Hague, R., Malik, W. U.: J. Indian Chem. Soc. *41*, 394 (1964)
81. Gillard, R. D., Laurie, S. H., Price, D. C., Phipps, D. A., Weick, C. F.: J.C.S. Dalton *1974*, 1385
82. Friedheim, C., Keller, F.: Ber. *39*, 4301 (1906)
83. Baker, L. C. W., Loev, B., McCutcheon, T. P.: J. Am. Chem. Soc. *72*, 2374 (1950)
84. Baker, L. C. W., Foster, G., Tan, W., Scholnick, F., McCutcheon, T. P.: J. Am. Chem. Soc. *77*, 2136 (1955)
85. Walfe, C. W., Block, M. L., Baker, L. C. W.: J. Am. Chem. Soc. *77*, 2200 (1955)
86. Shimura, Y., Ito, H., Tsuchida, R.: J. Chem. Soc. Jpn. *75*, 560 (1954)
87. Baker, L. C. W., McCutcheon, T. P.: J. Am. Chem. Soc. *78*, 4503 (1956)
88. Shimura, Y., Tsuchida, R.: Bull. Chem. Soc. Jpn. *30*, 502 (1957)
89. Baker, L. C. W., Baker, V., Eriks, K., Pope, M. T., Shibata, M., Pollins, O. W., Frang, J. H., Koh, L. L.: J. Am. Chem. Soc. *88*, 2329 (1966)
90. Shibata, M., Baker, L. C. W.: Abstracts of Papers, 138th National Meeting of the American Chemical Society, New York, N.Y., 1960
91. Evans, H. T., Jr., Showell, J. S.: J. Am. Chem. Soc. *91*, 6881 (1969)
92. Davies, G., Hung, Y. W.: Inorg. Chem. *15*, 704 (1976)
93. Davies, G., Hung, Y. W.: Inorg. Chem. *15*, 1358 (1976)
94. Tanner, S. P.: Inorg. Chem. *17*, 600 (1978)

3 Preparative Application of Chromatography

3.1 Introduction

The ion-exchange chromatography was first applied to the separation of geometrical isomers of a cobalt(III) complex by King and Walters [1], who separated *trans*- and *cis*-$[Co(NO_2)_2(NH_3)_4]^+$. From the cation-exchange resin (Amberlite IR-120), 1 mol/dm^3 NaCl eluted the *trans*-isomer and then the *cis*-isomer was eluted with 3 mol/dm^3 NaCl. This difference was attributed to that the *cis*-isomer was more firmly held because of its larger dipole moment.

Later, similar techniques were applied [2,3,4] to various complexes such as *trans*- and *cis*-$[Coa_2(en)_2]^+$ (a = NCS^-, NO_2^-, Cl^- and Br^-), *trans*- and *cis*-$[CoCl(NO_2)$-$(en)_2]^+$ and *trans*(NH_3)- and *cis*(NH_3)-$[CoCl_2(NH_3)_2(en)]^+$; from the elution curves it was confirmed that every *trans*-isomer is eluted faster than the corresponding *cis*-isomer.

Paper chromatography to known amminenitro-complexes $[Co^{III}(NO_2)_n(NH_3)_{6-n}]$ (n = 0, 1, 2, 3, 4, 6) was applied by Yamamoto et al. [5]. They found in four solvents that the R_F values increase almost regularly with the number of NO_2^- in the complex, and that *cis*-$[Co(NO_2)_2(NH_3)_4]^+$ exhibits a larger R_F value than the corresponding *trans*-isomer. With their procedure applied to the mother liquors from the syntheses of *mer*-$[Co(NO_2)_3(NH_3)_3]$ [6] and *trans*-$K[Co(NO_2)_4(NH_3)_2]$ [7] they detected *trans*-$[Co(NO_2)_2(NH_3)_4]^+$ in the former and *mer*-$[Co(NO_2)_3(NH_3)_3]$ in the latter synthesis as by-products. The same trend $R_F(cis) > R_F(trans)$ was reported by Stefanović and Janjić [8,9]. These instances show that the early workers were interested in applying chromatographic techniques to the separation of simple and classical complexes.

In the 1960's, when new methods for the preparation of complexes and spectroscopic methods for their physicochemical studies were concurrently progressing, coordination chemists noted applications of chromatographic techniques to the preparative studies of mixed ligand complexes. Ion-exchange column techniques, in particular, were most rapidly developed for the preparative studies.

Druding and Kauffman [10], who reviewed the chromatography of coordination complexes up to around 1967, described that "despite the rapid progress in the last two decades, a few problems remain to be solved, such as the separation of geometric isomers of high ionic charge and more efficient resolution of optical isomers." These problems have now been solved. Thus, the chromatographic techniques enabled the preparative separation of geometrical isomers, diastereoisomers, and enantiomeric isomers. Now chromatographic separations serve for the quantitative determination of the composition of reaction products.

46

In this chapter we shall review recent studies on preparative chromatography. The description is intended to note the preparative method for an aimed complex, the condition under which isomers of the complex are eluted, the elution order of the isomers, and the isomeric composition of the product.

3.2 Chromatography on Ion-Exchange Resins

3.2.1 Complexes with Multidentate Ligands

The practical uses of ion-exchange chromatography for the preparation of new complexes were demonstrated by Legg and Cooke [11]; two geometrical isomers of the cobalt(III) complex with ethylenediamine-N,N'-diacetic acid (H_2edda) and ethylenediamine, [Co(edda)(en)]$^+$ are *trans* and *cis* with respect to oxygens of the edda ligand (Fig. 3.1). Both isomers were separated on a cation exchange resin (Dowex 50W-X8, Na$^+$ form) by elution with 0.5 mol/dm^3 NaClO$_4$. The *trans* isomer was eluted earlier than the *cis* one, and much less *trans* isomer was formed in the reaction than *cis*. In the experiments with the corresponding complexes with the N-substituted analogues such as N,N'-dimethylethylenediamine-N,N'-diacetate and N,N'-diethylethylenediamine-N,N'-diacetate, no *cis* isomers were found, indicating the steric requirements of these ligands.

Similar studies were carried out with complexes of the formula [Co(dien)(L)]$^+$, where dien is diethylenetriamine and L is iminodiacetate (ida), methyliminodiacetate (mida), or pyridine-2,6-dicarboxylate (pdc) [12]. Each complex was prepared by the reaction of the ligand L with the [CoCl$_3$(dien)] complex in the presence of activated charcoal, and the resulting solution was chromatographed on a column of the same cation-exchange resin by using 0.5 mol/dm^3 NaClO$_4$ eluent.

In the case of [Co(ida)(dien)]$^+$, three isomers, *trans*(O), *s-cis*(O) and *u-cis*(O), were separated in that order of elution; the isomeric composition is given in Fig. 3.2. With [Co(mida)(dien)]$^+$, two isomers, *trans* and *s-cis*, were obtained, but the *u-cis* isomer was not found. The ligand pdc should only span meridional edges of an octahedron and only the *trans* isomer was found of [Co(pdc)(dien)]$^+$. In accordance with a general trend on *trans* and *cis* isomeric complexes, the *trans* isomers were first eluted.

The same workers prepared [Co(D-asp)(dien)]$^+$ (D-asp = D-aspartate) [13] by the reaction of [CoCl$_3$(dien)] and Ag$_2$D-asp at activated charcoal. The resulting solution was chromatographed on a cation exchanger column (Dowex 50W-X8, Na$^+$ form) by elution with 0.3 mol/dm^3 NaClO$_4$. The possible three isomers were isolated with different yields (Fig. 3.3).

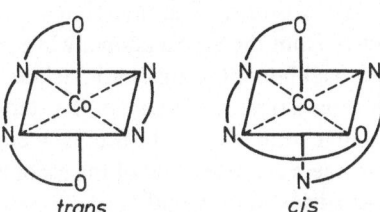

trans   cis   **Fig. 3.1.** Isomers of [Co(edda)(en)]$^+$

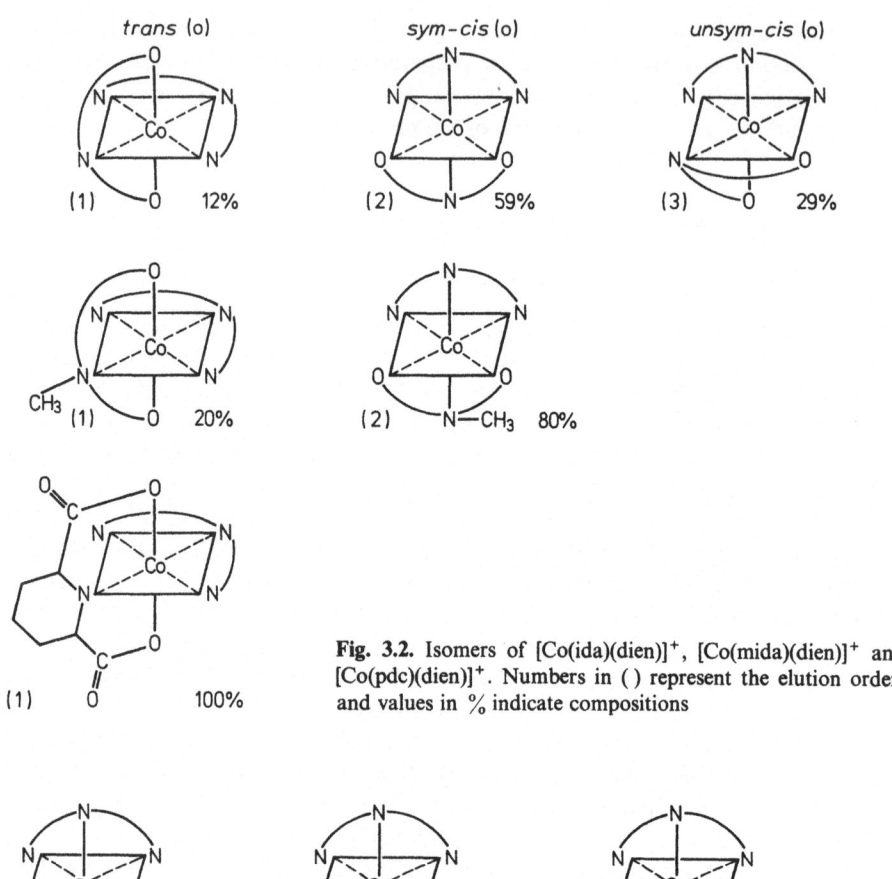

Fig. 3.2. Isomers of [Co(ida)(dien)]$^+$, [Co(mida)(dien)]$^+$ and [Co(pdc)(dien)]$^+$. Numbers in ( ) represent the elution order, and values in % indicate compositions

Fig. 3.3. Isomers of [Co(D-asp)(dien)]$^+$. Numbers in ( ) represent the elution order, and quantities in g indicate the yields

The four possible isomers for the complex with ethylenediamine-N,N'-di-L-α-propionate (LL-eddp) and ethylenediamine, [Co(LL-eddp)(en)]$^+$ (Fig. 3.4), were prepared [14] by the method of Legg and Cooke for the related edda complex. The resulting solution was chromatographed (Dowex 50W-X8, Na$^+$ form; 0.35 mol/dm^3 NaClO$_4$); two *trans* isomers were eluted before the *cis* isomer. The first eluted one was L-*trans* in which both methyl groups point away from the en backbone, while in the later eluted D-*trans* both point toward the en backbone. On this basis Schoenberg et al. [14] considered that the retention of the *trans* isomers arises from the steric interactions of the methyl groups with the ion-exchange resin. One of the four isomers, D-*cis*, was not detected and the yields of the *trans* isomers exceeded that of the *cis* one, indicating preferential *trans* orientation of this quadridentate ligand.

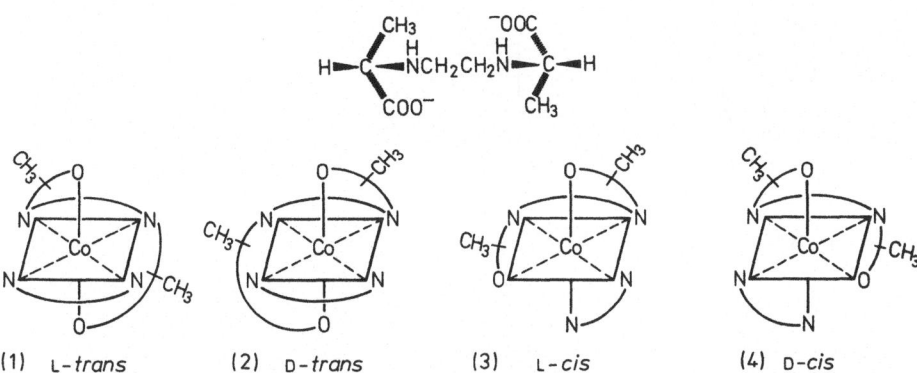

Fig. 3.4. LL-eddp and isomers of [Co(LL-eddp)(en)]$^+$ (from Ref. [14]). Numbers in ( ) represent the order of elution

For the bis(L-aspartato)cobaltate(III) ion, [Co(L-asp)$_2$]$^-$, three geometrical forms, *trans*(N), *cis*(N)*trans*(O$_5$O) and *cis*(N)*trans*(O$_6$O), are possible (Fig. 3.5). Hosaka et al. [15] isolated two *cis*(N) isomers by ion-exchange chromatography from a reaction mixture of [Co(CO$_3$)$_3$]$^{3-}$ and L-aspartate. Yamada et al. [16] isolated all the possible isomers from a reaction mixture of freshly prepared CoO(OH) and Na$_2$L-asp with activated charcoal by column chromatography (Dowex 1-X2, Cl$^-$ form; 0.1 mol/dm^3 NaCl eluent). An analogous complex, [Co(L-asp)(ida)]$^-$, is also separated into three isomers, *trans*(N), *cis*(N)*trans*(O$_5$O) and *cis*(N)*trans*(O$_6$O), which are shown in Fig. 3.5, (Dowex 1-X4, Cl$^-$ form; 0.1 mol/dm^3 LiCl). [17]

The sexadentate ligand, (L)-ethylenediamine-*N,N*-diacetic-*N'*-monosuccinic acid (H$_4$eddams), can form two geometric isomers with cobalt(III) (Fig. 3.6), in which

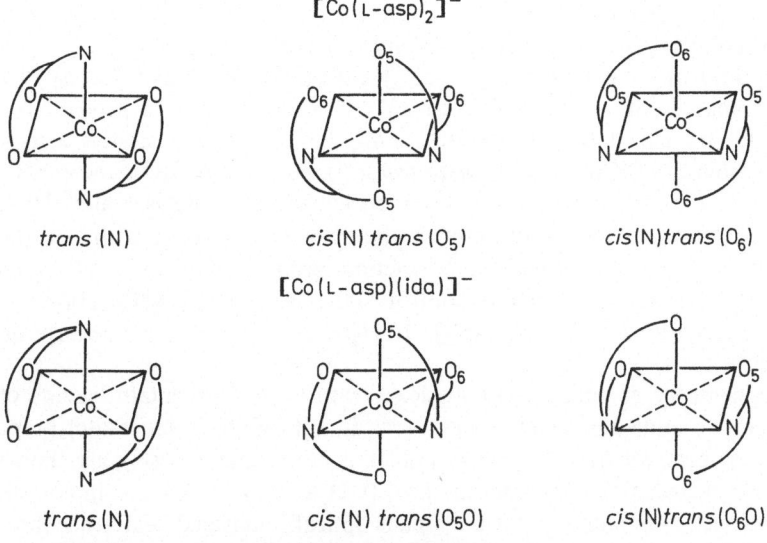

Fig. 3.5. Isomers of [Co(L-asp)$_2$]$^+$ and [Co(L-asp)(ida)]$^-$

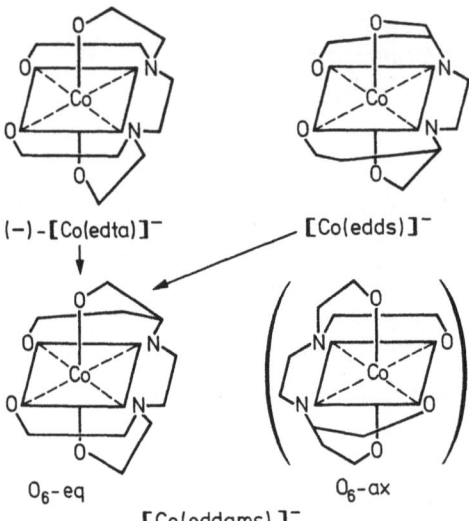

(−)-[Co(edta)]⁻

[Co(edds)]⁻

O₆-eq

O₆-ax

[Co(eddams)]⁻

Fig. 3.6. Possible isomers for [Co(eddams)]⁻ and their relation ship to [Co(edta)]⁻ and [Co(edds)]⁻ (from Ref. [18])

O_6-eq is the isomer containing a six-membered chelate ring in equatorial coordination, O_6-ax containing the ring in axial coordination. Legg and Neal [18] prepared this complex in order to investigate the absolute configuration of aspartic acid complexes by CD spectroscopy; the crude ligand, containing other closely related sexadentate ligands such as (L,L)-N,N'-ethylenediaminedisuccinic acid (H$_4$edds) and ethylenediaminetetraacetic acid (H$_4$edta), was allowed to react with a suspension of Na$_3$[Co(CO$_3$)$_3$] · 3 H$_2$O in water in the presence of activated charcoal, and the resulting solution was chromatographed (Dowex 1-X8, Cl⁻ form). By elution with 0.1 mol/dm^3 NaCl, the desired complex [Co(eddams)]⁻ was separated from the related complexes, [Co(edds)]⁻ and [Co(edta)]⁻. The complex isolated was only O_6-eq isomer, exhibiting absolute stereospecificity of the eddams ligand.

The CD spectra of the complexes, [Co(ida)$_2$]⁻, [Co(ida)(asp)]⁻ and [Co(asp)$_2$]⁻, should be compared with those of [Co(edta)]⁻, [Co(eddams)]⁻ and [Co(edds)]⁻, respectively, for the elucidation of absolute configurations. Their tentative assignments for the isomers of [Co(L-asp)$_2$]⁻ were *trans*(N), *cis*(N)*trans*(O$_5$O), and *cis*(N)-*trans*(O$_6$O), in the order of elution, opposite to those made by Douglas et al. [17,19] as to the two *cis*(N) isomers. An X-ray study was made by Oonishi et al. with a crystal of calcium bis(aspartato)cobaltate(III), a reaction product of [Co(CO$_3$)$_3$]$^{3-}$ and L-aspartate [20,21]. The structural analysis confirmed the formula Ca *cis*(N)*trans*(O$_5$O)-[Co(L-asp)$_2$] · *cis*(N)*trans*(O$_6$O)-[Co(L-asp)$_2$] · 10 H$_2$O⁻ assigned from the absorption and CD spectra.

The stereochemistry of cobalt(III) complexes with O,N,O-terdentate ligands of the linear type was studied by Okamoto et al. [22]. In a [Co(O,N,O-terdentate)$_2$]-type complex four isomers, *fac-trans*(N), *mer-trans*(N), Λ-*cis*(N) and Δ-*cis*(N), are shown in Fig. 3.7. When L-alanine-N-monoacetic acid (L-H$_2$alama) participates in coordination, the nitrogen donor atom of the L-alama is optically activated, and hence three isomers, RR, RS, and SS, are possible for each of the above *trans*(N) and *cis*(N) forms.

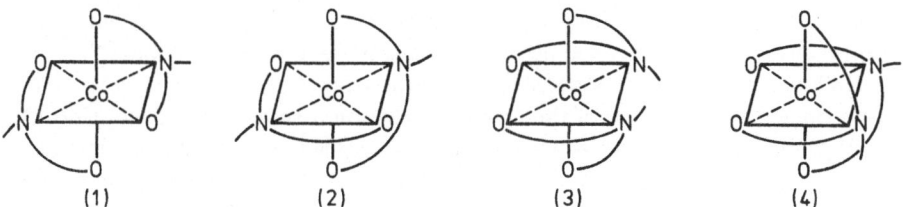

Fig. 3.7. Four possible isomers for a [Co(O, N, O)₂]-type complex: (1) *facial-trans*(N), (2) *meridional-trans*(N), (3) Λ-*cis*(N) and (4) Δ-*cis*(N) (from Ref. [22]).

Experimentally, *trans*(N)-RR, *trans*(N)-RS, Δ-*cis*(N)-RR, Δ-*cis*(N)-RS, Λ-*cis*(N)-RR and Λ-*cis*(N)-RS were chromatographically isolated from a reaction mixture obtained after lead dioxide oxidation (Dowex 1-X8, Cl⁻ form; 0.07 mol/dm³ KCl or NaClO₄). The existence of *mer-trans*(N) isomers was excluded because of the general difficulty of meridional coordination of a linear O,N,O-ligand.

With the corresponding L-prolinato-*N*-monoacetato complex, [Co(L-proma)₂], only a *trans* isomer and a *cis* isomer were detected; this is explained by a unique R configuration of the nitrogen donor atom and by the preferential facial coordination of the L-proma ligand [23]. Similar experiments were carried out with [Co(L-promp)₂]⁻, [Co(L-proma)(L-promp)]⁻, [Co(L-proma)(ida)]⁻ and [Co(L-promp)(ida)]⁻ (L-promp = L-proline-*N*-monopropionate), and it was found that all the isomers obtained in solid state have *trans*(N)-RR or *trans*(N)-R configuration.

For the [Co(edda)(diamine)]⁺ system, two geometries, *s-cis* and *u-cis*, are possible; when the diamine is an unsymmetrical ligand such as propylenediamine (pn) additional isomerism is possible for the *u-cis* geometry. Figure 3.8 shows four *u-cis*-[Co(edda)-(R-pn)] isomers distinguished by the *cis-cis*(N,O) (the nitrogen next to the optically active carbon of R-pn is *cis* to both chelated edda oxygen atoms) and the *cis-trans*-(N,O) isomers (the same R-pn nitrogen is *cis* to one oxygen and *trans* to the other).

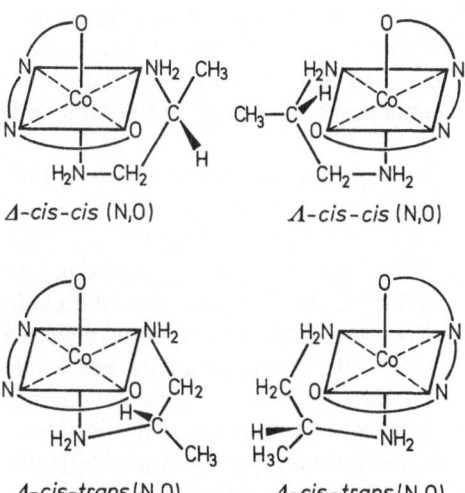

Δ-*cis-cis* (N,O) Λ-*cis-cis* (N,O)

Δ-*cis-trans* (N,O) Λ-*cis-trans* (N,O) **Fig. 3.8.** Isomers of *u-cis*-[Co(edda)(R-pn)]⁺

The thermodynamically favored *s-cis* isomers were well characterized but the *u-cis* isomers were not characterized because of being present in trace amount [24].

Halloran and Legg [25] succeeded in synthesizing sufficient quantities of these *u-cis* isomers; the reaction of the ligand H_2edda on a suspension of $Na_3[Co(CO_3)_3]$ · 3 H_2O in water without activated charcoal yielded an equilibrium mixture of *s-cis-* and *u-cis*-$[Co(CO_3)(edda)]^+$. The carbonato complex was allowed to react with (R)-propylenediamine dihydrochloride and the resulting solution was purified on a column of a cation exchange resin (Dowex 50W-X8, Na^+ form). NaH_2PO_4- Na_2HPO_4 (buffer 0.2 mol/dm³, pH 6.8) was efficient for the separation. The ion-exchange chromatography is indispensable for the isolation of the *u-cis* isomers.

With respect to related complexes $[Co(medds)]^-$ and $[Co(dmedds)]^-$ (medds = (S,S)-*N*-methylethylenediaminedisuccinate and dmedds = (S,S)-*N,N'*-dimethyl-ethylenediaminedisuccinate), the absolute stereospecificity in their formation was found by Jordan and Legg [26]; an aqueous solution containing cobalt(II) salt and a mixture of both ligands was aerated over activated charcoal; the resulting solution was treated with a Sephadex G-15 column (Cl⁻ form) and chloroform-saturated water to separate sodium chloride from the complex anions. The effluent was then chromatographed (Dowex 1-X8, Cl⁻ form; 0.0064 mol/dm³ NaCl eluent). Three bands corresponding to $[Co(dmedds)]^-$, $[Co(medds)]^-$ and $[Co(edds)]^-$ were eluted in that order. The complexes isolated showed only slight differences in their CD spectra, indicating the same absolute configuration and also a very small contribution of the asymmetric nitrogen atoms to rotational strength in the EDDS system.

3.2.2 Complexes with Amino Acid and Diamine or Oxalic Acid

Matsuoka et al. [27~30] isolated and characterized geometrical isomers of $[Co(am)_2$-$(en)]^+$ and $[Co(am)_2(ox)]^-$ type complexes (am = amino-acidate) in order to investigate a "vicinal effect" displayed by the coordinated optically active α-amino-acidate.

The ethylenediaminebis(glycinato) complex $[Co(gly)_2(en)]^+$ was prepared by oxidizing an aqueous solution of the components with lead dioxide; the resulting solution was chromatographed (Dowex 50W-X8, H^+ form) [27] with 0.5 mol/dm³ KBr to separate three possible isomers, *trans*(N), *cis*(N)*cis*(O) and *cis*(N)*trans*(O) with respect to N and O of chelated glycinate ions, in that order of elution. Similarly, three possible isomers of the $[Co(ox)(gly)_2]^-$ complex were separated in the same order (Dowex 1-X10, Cl⁻ form; 0.5 mol/dm³ $NaClO_4$) [27].

Similar experiments were then carried out with optically active amino acids. With the oxalatobis(L-serinato)cobaltate(III) complex, $[Co(ox)(L-ser)_2]^-$, all possible isomers were eluted in the order of $\Delta(-)_{546}$-*trans*(N), $\Lambda(+)_{546}$-*trans*(N), $\Delta(-)_{546}$-*cis*(N)-*trans*(O), $\Lambda(+)_{546}$-*cis*(N)*cis*(O), $\Delta(-)_{546}$-*cis*(N)*cis*(O) and $\Lambda(+)_{546}$-*cis*(N)*trans*(O) (Dowex 1-X10, Cl⁻ form; 0.07, 0.27 mol/dm³ KCl) [28]. Concerning the elution order of the four *cis*(N) isomers, the workers stated that according to molecular models a hydrogen-bond interaction between —OH of one L-serinate and —NH_2 of another L-serinate in a complex decreases in the order parallel to the elution order.

All isomers of $[Co(L-ser)_2(en)]^+$ were also separated in the following order: $\Lambda(+)_{589}$-*trans*(O), $\Delta(-)_{589}$-*trans*(O), $\Lambda(+)_{589}$-*cis*(O)*trans*(N), $\Delta(-)_{589}$-*cis*(O)*cis*(N), $\Lambda(+)_{589}$-*cis*(O)*cis*(N) and $\Delta(-)_{589}$-*cis*(O)*trans*(N) (Dowex 50W-X8, H^+ form; suc-

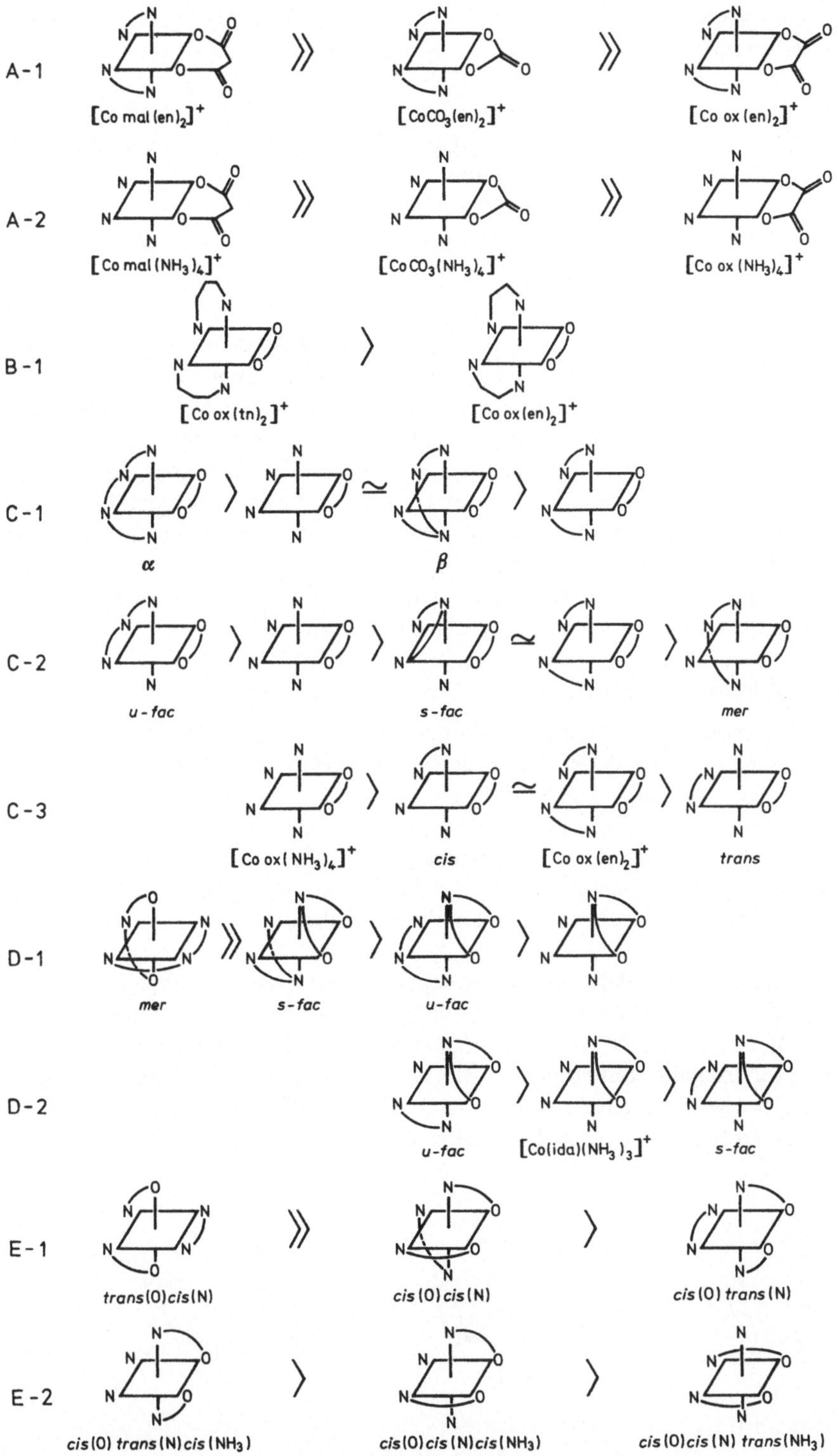

Fig. 3.9. Elution orders for $[Co(N)_4(O)_2]^+$ complexes (from Ref. [31])

cessive elutions with 0.05, 0.1, 0.2 and 0.5 mol/dm³ NaClO₄)²⁹⁾. With the bis-(L-alaninato)ethylenediamine complex, [Co(L-ala)₂(en)]⁺, the elution order was found to be $\Delta(-)_{589}$-*trans*(O), $\Lambda(+)_{589}$-*trans*(O), $\Delta(-)_{546}$-*cis*(O)*cis*(N), $\Delta(-)_{546}$-*cis*(O)-*trans*(N), $\Lambda(+)_{546}$-*cis*(O)*cis*(N) and $\Lambda(+)_{546}$-*cis*(O)*trans*(N) (Dowex 50W-X8, H⁺ form; 0.02 mol/dm³ NaClO₄)³⁰⁾. With [Co(L-ala)₂(ox)]⁻, the isomers were isolated in the following order: $\Lambda(-)_{546}$-*trans*(N), $\Delta(+)_{546}$-*trans*(N), $\Lambda(-)_{546}$-*cis*(N)*cis*(O), $\Delta(+)_{546}$-*cis*(N)*cis*(O), $\Delta(+)_{546}$-*cis*(N)*trans*(O) and $\Delta(-)_{546}$-*cis*(N)*trans*(O) (Dowex 1-X8, Cl⁻ form; 0.02~0.01 mol/dm³ KCl)³⁰⁾.

3.2.3 Elution Order in the [Co(N)₄(O)₂]⁺-Type Complexes

The column chromatography using ion-exchange resin is useful, in particular, for univalent cations with the [Co(N)₄(O)₂]⁺ chromophore or anions with the [Co(N)₂-(O)₄]⁻ chromophore. With complexes of the [Co(N)₄(O)₂]⁺ type, Kobayashi and Shibata³¹⁾ examined the order of elution. They used columns containing 100~200 mesh Dowex 50W-X8 in Na⁺ form (∅ 2.0~7.0 cm, 10~50 cm long), and eluted with 0.3~0.4 mol/dm³ NaCl in rates of 0.3~1.2 cm³/min. The complexes were classified into several series (A~E) each of which was chromatographed. The results are summarized in Fig. 3.9, where the > notation indicates that a complex species written on the left side is eluted earlier than that on the right side (≫ indicates a large separation, and ≃ means nearly coincident elution).

3.2.4 Non-Electrolyte Complexes

The three isomers for diamminecarbonatoglycinatocobalt(III) [CoCO₃(gly)(NH₃)₂] are represented as *mer*(*cis*), *mer*(*trans*) and *fac* isomers (Fig. 3.10). Kanazawa and Shibata³²⁾ separated these isomers on a cation exchange column (Dowex 50W-X8, Na⁺ form); the complex was prepared by the reaction of [Co(NH₃)₂(CO₃)₂]⁻ and glycine and the resulting solution was charged on the column. With water, the negatively charged complexes such as the starting material and [Co(CO₃)(gly)₂]⁻ effused first, and then the desired complex effused in the order of *mer*(*cis*), *mer*(*trans*) and *fac*.

This technique was widely used in the separation of isomeric complexes of non-electrolyte type³³⁾³⁴⁾³⁵⁾. For the L-aspartatoethylenediamineoxalato complex [Co(ox)-(L-asp)(en)]⁻, four isomers, *mer*-Λ, *mer*-Δ, *fac*-Λ and *fac*-Δ, are possible (see Fig. 4.3). With an anion-exchange resin (Dowex 1-X8, Cl⁻ form) an extremely long column was necessary since washing gave a diffuse band on the column and successive elution with an electrolyte made each band spread considerably. In contrast, complete

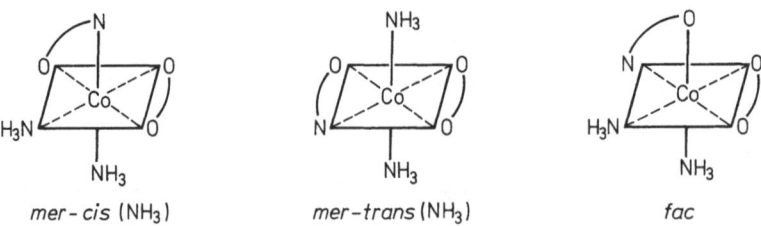

$mer-cis$ (NH₃) $mer-trans$ (NH₃) fac

Fig. 3.10. Possible geometrical isomers for [Co(CO₃)(gly)(NH₃)₂]

separation of these isomers was achieved at a cation exchange resin in H$^+$ form (Dowex 50W-X8) with water alone, where the isomers to be separated behaved as neutral species [Co(ox)(L-Hasp)(en)]. The two-week elution on an anion-exchange column was reduced to two days on a cation-exchange column [36].

A cation-exchange column (Dowex 50W-X4, Na$^+$ form) [25] separates the *fac* and *mer* isomers of the neutral [Co(L-ala)(edda)] complex by elution with water.

For the mixed ligand complex with L- or D-aspartic acid and L-2,4-diaminobutyric acid, [Co(L- or D-asp)(L-2,4-dba)], the possible isomers are shown in Fig. 3.11. Both of L- and D-series are composed of two meridional forms and one facial form about the [Co(N)$_3$(O)$_3$] chromophore. The two L-*mer* isomers are denoted by L-*transO$_5$-cisN$_5$* (with respect to oxygen) and L-*cisO$_5$transN$_5$*. To investigate the ^{1}H NMR spectra of those isomers, Watabe et al. [37] prepared the complexes with Na$_3$[Co(CO$_3$)$_3$] · 3 H$_2$O and also by the direct synthesis from the components. The reaction mixture was first subjected to an anion exchange resin (Dowex 21k, OH$^-$ form) and then to a column of cation-exchange resin (Dowex 50W-X8, Na$^+$ form; H$_2$O eluent). Three isomers, L-*transO$_5$cisN$_5$*, L-*cisO$_5$transN$_5$* and L-*fac* of the L-aspartato complex, two, D-*cisO$_5$-cisN$_5$* and D-*fac*, of the D-aspartato complex were isolated.

From this study, Watabe et al. found a general trend that the methine proton in a H—C—N—Co(III)—X fragment, where X is a nitrogen or oxygen atom occupying the *trans* position to N, resonates at a higher or lower field according to whether X is oxygen or nitrogen. The trend of the chemical shifts holds for other complexes such as [Co(L-asp)$_2$]$^-$ [19], [Co(L-asp)(ida)]$^-$ [17] and [Co(L-2,4-dba)$_2$]$^+$. This "through cobalt effect" has been theoretically rationalized by Yoneda et al. [38]. The remaining third isomer, D-*transO$_5$transN$_5$*, was found in the isomerization product of the known two isomers in the presence of activated charcoal [30].

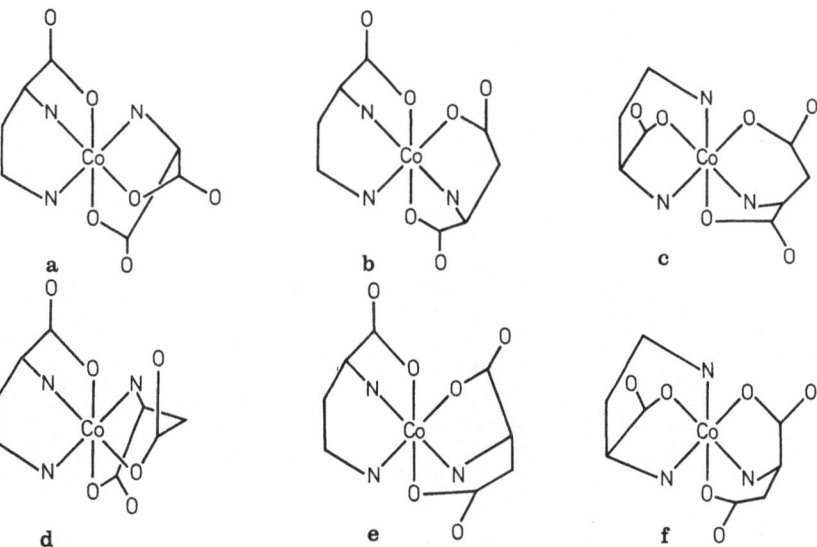

Fig. 3.11a–f. Possible geometrical isomers of [Co(L- or D-asp) (L-2,4-dba)] (from Ref. [37]). **a** D-*cisO$_5$-cisN$_5$*. **b** D-*transO$_5$transN$_5$*. **c** D-*fac*. **d** L-*transO$_5$cisN$_5$*. **e** L-*cisO$_5$transN$_5$*. **f** L-*fac*

3.3 Chromatography on Ion-Exchange Cellulose or Ion-Exchange Dextran

3.3.1 Tris(Diamine) Complexes

In 1966, [Co(en)₃] Br₃ was separated into its optically active enantiomers [40] on a column of anion-exchange resin, which was in advance loaded with tartrate or antimonyltartrate ions. The resolution was partial. In the same year, Brubaker et al. [41] achieved the total resolution of a trinuclear cobalt(III) complex, hexakis(2-amino-ethanethiolato)tricobalt(III) bromide on a column of a cation-exchange cellulose (Bio-Rad Cellex CM) by eluting with 0.1 mol/dm³ NaCl.

After several attempts with unsufficient resolution [42], total resolution [43] was achieved for [Co(en)₃] Br₃ dissolved in water on a column (1.1 × 120 cm) filled with an ion-exchange dextran, SE-Sephadex[13] C-25 in Na⁺ form. The elution with 0.15 mol/ dm³ sodium (+)-tartrate showed two completely separated peaks corresponding to the Δ(−)₅₈₉- and Λ(+)₅₈₉-isomer, respectively. Since then column chromatography on Sephadex became a powerful tool for the resolution of various complexes or for the separation of geometrical isomers and diastereoisomers.

Yoshikawa and Yamasaki have reviewed the resolution of optical isomers and the separation of geometrical isomers of cobalt(III) complexes on Sephadex ion-exchangers [44], including the experimental techniques.

For the tris[(±)-propylenediamine]cobalt(III) ion, there are 24 possible isomers (Table 3.1, where symbols Δ and Λ refer to the absolute configurations of the complex ion, *lel* and *ob* to the conformations of the chelate ring, and *fac* and *mer* to the isomerism with respect to the methyl groups of the chelated pn molecules). Dwyer et al. [45] studied this system by column chromatography with cellulose powder and paper chromatography (see Section 3.4). MacDermott [46] separated the *fac* isomer from the *mer* one by the fractional crystallization of the Δ(−−−) isomer [45]. The *fac* isomer was isolated as crystalline bromide and the *mer* isomer as amorphous glasses, the ratio being 1:3.

Table 3.1. Possible Isomers for [Co{(±)-pn}₃]³⁺

Conformation	optical isomers	isomers with respect to CH₃ groups	total isomers
lel_3	Λ (+++)	*fac, mer*	2
	Δ (−−−)	*fac, mer*	2
lel_2ob	Λ (++−)	*fac, mer* (3)	4
	Δ (−−+)	*fac, mer* (3)	4
$lel\ ob_2$	Λ (+−−)	*fac, mer* (3)	4
	Δ (−++)	*fac, mer* (3)	4
ob_3	Λ (−−−)	*fac, mer*	2
	Δ (+++)	*fac, mer*	2

[13] For cation-exchanging Sephadex, SE-, SP- and CM-sephadex, having sulfoethyl, sulfopropyl, and carboxymethyl groups respectively, are available. SE-Sephadex can be replaced by SP-Sephadex; both have almost the same physical and chemical properties.

Kojima et al. [47)] isolated both the *fac* and *mer* isomers of $\Lambda(-)_{589}[\text{Co}(\text{R}(-)\text{-pn}_3)]^{3+}$ in pure state by Sephadex chromatography; a solution resulting from the reaction of $[\text{CoBr}(\text{NH}_3)_5]\,\text{Br}_2$ and (R)-propylenediamine with activated charcoal was subjected to an SP-Sephadex column $(2.7 \times 135\text{ cm})$. By elution with $0.18\text{ mol/dm}^3\ \text{Na}_2\text{SO}_4$, the $\Delta(---)$ and $\Lambda(---)$ isomers were completely separated in this order. The fractions of the $\Lambda(---)$ isomer were re-eluted from a SP-Sephadex column with 0.15 mol/dm^3 sodium (+)-tartratoantimonate(III) [14] and completely separated into the fast moving *mer* isomer and the slow moving *fac* isomer, the ratio being *mer*:*fac* = 3:1.

Harnung et al. [48)] studied the $[\text{Co}(\text{R,S-pn})_3]^{3+}$ system in detail. A mixture of $\Lambda(-)_{589}[\text{Co}(\text{R,S-pn})_3]\,\text{Cl}_3$, R,S-pn · 2 HCl and R-pn · 2 HCl in water was refluxed over activated charcoal at 100 °C. The resulting equilibrium mixture of isomers was chromatographed on a SP-Sephadex column. By elution with $0.1\text{ mol/dm}^3\ \text{Na}_3\text{PO}_4$, four racemic fractions, *lel*₃, *lel₂ob*, *ob₂lel* and *ob*₃, appeared in this order. Every fraction was rechromatographed (SP-Sephadex C-25; 0.15 mol/dm^3 (+)-tartrate), the Λ species being eluted earlier than the Δ species. For the $\Lambda(lel_3)$ and $\Delta(lel_3)$ isomers, columns of $2.5 \times 100\text{ cm}$ and a $0.1\text{ mol/dm}^3\ \text{Na}_3\text{PO}_4$ eluent were used. The isomer distributions are cited in Table 3.2, where F represents the molar ratio of S-pn/R-pn, and the calculated compositions are the values based upon the experimental values in the first column.

The other examples are $[\text{Co}\{(-)\text{cptn}\}_3]^{3+}$ [49)], $[\text{Co}(\text{dabp})(\text{R-pn})_2]^{3+}$ [50)], $[\text{Co}(\text{ibn})_3]^{3+}$ [51)], $[\text{Co}(\text{tmd})_3]^{3+}$ [52)], $[\text{Co}(\text{S-bn})_3]^{3+}$ [53)], $[\text{Co}(\text{en})_x(\text{tn})_y(\text{tmd})_z]^{3+}$ [54)], etc. ((−)cptn = = $(-)_{589}$-*trans*-1,2-diaminocyclopentane, dabp = 2,2′-diaminobiphenyl, ibn = = 2-methyl-1,2-propanediamine, tmd = tetramethylenediamine, S-bn = (S)-1,3-butanediamine).

Table 3.2. Equilibrium Isomer Distribution at 100 °C

Isomer	Experimental composition %			Calculated composition %
	$F = 1.00$	$F = 1.90$		$F = 1.90$
Λlel_3	35.0	37.9	37.9	38.0
Δlel_3		5.6	5.9	5.5
Λlel_2ob	41.1	23.6	22.4	23.5
Δlel_2ob		12.4	11.9	12.4
Λob_2lel	18.0	5.1	5.1	5.4
Δob_2lel		10.2	10.0	10.3
Λob_3	4.0	—	—	0.6
Δob_3		3.9	4.0	4.3
Total	98.1	98.7		100.0

---

[14] Preparation: Diantimony trioxide (462 g) is slowly added to an aqueous solution of sodium hydrogentartrate monohydrate $\text{NaC}_4\text{H}_5\text{O}_6 \cdot \text{H}_2\text{O}$ (600 g) in 1.2 dm^3 of H_2O), with stirring. The mixture is stirred for 2 h at 80–90 °C. After cooling, the solution is filtered and used as the eluent after suitable dilution with water. When a large amount of ethanol is added to the filtrate, crystals of $\text{Na}_2[\text{Sb}_2(\text{C}_4\text{H}_2\text{O}_6)_2] \cdot 5\text{ H}_2\text{O}$ precipitate. The aqueous solution is acidic.

In order to prepare optically active Sephadex cation exchanger with (—)-tartrate residues, Fujita et al. [55] prepared two types of exchangers, one having the residues bound to the Sephadex base as ether (left), another as ester (right):

$$\text{Sephadex}-O-\underset{\underset{\text{NaOOC}}{|}}{\overset{\overset{\text{H}}{|}}{C}}-\underset{\underset{\text{H}}{|}}{\overset{\overset{\text{COONa}}{|}}{C}}-OH \qquad \text{Sephadex}-\ominus OC-\underset{\underset{\text{H}}{|}}{\overset{\overset{\text{HO}}{|}}{C}}-\underset{\underset{\text{OH}}{|}}{\overset{\overset{\text{H}}{|}}{C}}-COONa$$

They achieved total resolution on a column of the Sephadex cation exchanger with (—)-tartrate ester by elution with a (+)-tartrate solution [56]. Similarly, the resolution of [Co(tn)$_3$]$^{3+}$ was accomplished [57].

3.3.2 Polyamine Containing Complexes

For the bis-complex of diethylenetriamine, [Co(dien)$_2$]$^{3+}$, three topological isomers, *s-fac* (trans-*fac*), *u-fac* (cis-*fac*) and *mer*, are shown in Fig. 3.12. In addition, the *u-fac* and *mer* isomers are resolvable [58~60]. The complex containing three isomers was prepared by different methods [59]; the reaction of [Co(CO$_3$)$_3$]$^{3-}$ and dien in the presence of activated charcoal (**A**), the reaction of [CoBr(NH$_3$)$_5$]$^{2+}$ and dien in the presence of activated charcoal (**B**), aerial oxidation of a Co^{2+}—dien mixture in the presence of activated charcoal (**C**), and the reaction of *mer*-[CoCl$_3$(dien)] and dien in the absence of activated charcoal (**D**). From a SP-Sephadex column (Na$^+$ form) with 0.15 mol/dm^3 sodium (+)-tartrate, the fastest moving species was *s-fac*, the middle one *u-fac*, and the slowest one *mer* isomer. The formation ratio of *s-fac*:*u-fac*:*mer* was found to be 16:25:59 for method **A**, and 7:30:63 for method **B** and **C**. The method **D** produced only *u-fac*:*mer* = 8:92.

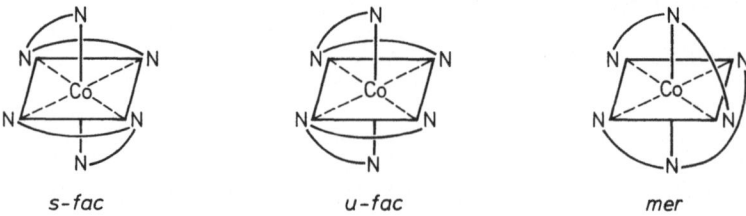

s-fac u-fac mer

Fig. 3.12. Topological isomers of [Co(dien)$_2$]$^{3+}$

These three isomers were separated on a column of P-cellulose, too. When 1-butanol saturated with water served as the solvent, 1-butanol:conc. HCl:water (200:15:15) eluted *mer*, *s-fac* and *u-fac* in this order. The resolutions of the *u-fac* and *mer* isomers were completely achieved on SE-Sephadex with 0.15 mol/dm^3 sodium (+)-tartrato-antimonate(III), the (+) and (—) enantiomers being eluted in this order. Since the *mer* isomer racemizes quickly in neutral or alkaline solution, all the operations were performed in an aqueous solution acidified with 0.01 mol/dm^3 HCl.

By chromatographic separation, Keene and Searle [61] determined equilibrium distributions of the three geometrical isomers of the complex prepared by aerial oxidation over activated charcoal. The results are cited, in part, in Table 3.3.

Table 3.3. Equilibrium Geometric Isomer Proportions

$(M = mol/dm^3, 18 °C)$

Solvent	Counterion	[Co], M	Added salt	Isomer proportions %		
				s-fac	u-fac	mer
H_2O	Cl^-	0.4, 0.2, 0.02		7	28	65
H_2O	Cl^-	0.02		12	41	47
H_2O	ClO_4^-	0.2, 0.02		8	30	62
DMSO	ClO_4^-	0.01		6	14	80
H_2O	Cl^-	0.02	NaOH (0.01 M) pH ~10	10	28	62
H_2O	Cl^-	0.02	Na_2PO_4 (0.4 M)	59	29	12
H_2O	Cl^-	0.02	Na_2SeO_3 (2 M)	59	28	13
H_2O	SO_4^{2-}	0.2, 0.02	K_2SO_4 (2 M)	25	30	37

The environmental parameters such as solvent, counterion, added salt and pH more or less affect the isomer proportions. Under the usual preparative conditions at 18 °C, the proportions are $s\text{-}fac:u\text{-}fac:mer = 7:28:65$ for aqueous solutions with Cl^- or Br^- as anion, indicating the preference of meridional coordination of dien over facial coordination. On the other hand, the addition of an oxy anion, PO_4^{3-} or SeO_3^{2-}, increases the proportion of the s-fac isomer at the expense of the mer one, the proportion of the u-fac remaining nearly constant. The enhanced proportion of the s-fac is ascribed to the differential specific associations between the oxy anion and the three isomeric cations, the interaction magnitudes being in the order $s\text{-}fac > u\text{-}fac > mer$. This interpretation is similar to the $\Delta\text{-}[Co(en)_3]^{3+} - PO_4^{3-}$ ion pair model proposed by Mason and Norman [62].

Closely analogous bis[di(2-aminoethyl)sulfide] complexes $[Co(daes)_2]^{3+}$ (daes = = $H_2NCH_2CH_2SCH_2CH_2NH_2$), were searched by Searle and Larsen [63] for three possible isomers; the complex ions prepared by aerial oxidation without or with phosphate, from $[Co(NH_3)_6]^{3+}$, from trans-$[CoCl_2(py)_4]Cl \cdot 6 H_2O$ in methanol or in pyridine, and from $Na_3[Co(CO_3)_3] \cdot 3 H_2O$. The chromatographic separation was carried out on a column of SP-Sephadex C-25 using a 0.15 mol/dm³ sodium (+)-tartratoantimonate(III) eluent. The product in each case was eluted in a single band, and the eluted fractions exhibited a partial resolution within the band. Complete resolution of the complex was achieved by the less soluble diastereoisomer formation with sodium (+)-tartratoarsenate(III).

The CD spectrum of the $(-)_{589}$-isomer showed a close correspondence to that of $(-)_{589}\text{-}u\text{-}fac\text{-}[Co(dien)_2]^{3+}$ over the ligand field and charge transfer regions; on this basis the $(-)_{589}[Co(daes)_2]^{3+}$ isomer was assigned to u-fac with the same absolute configuration as the reference dien isomer. Thus, only one isomer could be detected in the $[Co(daes)_2]$ system. In a similar manner, only one isomer, s-fac, was detected in the $[Co(dema)_2]^{3+}$ system (dema = N-methylbis(2-aminoethyl)amine or 4-methyl-diethylenetriamine, $H_2NCH_2CH_2N(CH_3)CH_2CH_2NH_2$). A conformation analysis indicated that this isomer is the most stable one among the three possible isomers [64].

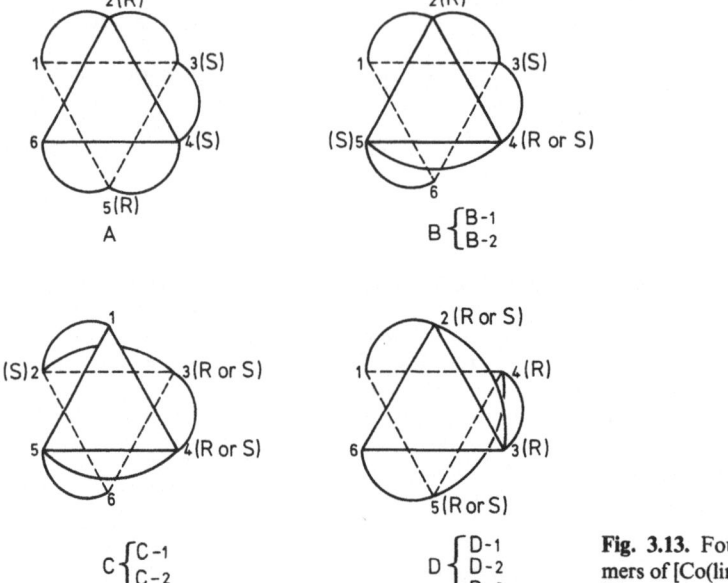

Fig. 3.13. Four geometrical isomers of [Co(linpen)]³⁺ (from Ref. [65])

When linear pentaethylenehexamine(1,14-diamino-3,6,9,12-tetraazatetradecane, linpen) coordinates to a cobalt(III) ion, there are four possible geometrical isomers, which are denoted by **A**, **B**, **C** and **D** in Fig. 3.13. If the absolute configurations (R and S) of secondary amine-N atoms are considered, eight isomers can be counted. As to **A**, the absolute configurations are fixed to R for N2 and N5, and S for N3 and N4, then only one isomer is possible. As to **B**, the absolute configuration of N4 can take either R or S, and hence two isomers are possible. As to **C**, both N3 and N4 can take the same absolute configuration, RR or SS, and hence two isomers are possible. For **D**, the absolute configuration of N2 is indenpendent of that of the N5, and consequently three isomers, RR, RS and SS, are possible [65]. The complex was prepared by the reaction of linpen with [CoBr(NH₃)₅]Br₂ at room temperature in the presence of activated charcoal. The resulting solution was chromatographed from SP-Sephadex. The procedures were repeated to attain fine

Table 3.4. Assigned Structures of Isomers

Isomer	Structure
(+)₅₈₉-I	**A**
(+)₅₈₉-II	**B1 + B2**
(+)₅₈₉-III	**A1 or C2**
(+)₅₈₉-IV	**C2 or C1**
(−)₅₈₉-V	**D1(SS)**
(−)₅₈₉-VI	**D2(RS)**
(−)₅₈₉-VII	**D3(RR)**

separations. From the final fractions, the desired isomers, I ∼ VII, were crystallized as the chlorides, [Co(linpen)]Cl$_3$ · n H$_2$O. The structures assigned on the basis of the absorption, CD, IR and PMR spectra are cited in Table 3.4.

The other examples are [Co(penten)]$^{3+}$ [66], [Co(R-mepenten)]$^{3+}$ [66], [Co(ida)-(dien)]$^+$ [67], u-fac-[Co(dien)(dema)]$^{3+}$, mer-[Co(dien)(dema)]$^{3+}$ [68] etc. (penten = = N,N,N',N'-tetrakis(2-aminoethyl)ethylenediamine, mepenten = N,N,N',N'-tetrakis(2-aminoethyl)-1,2-propanediamine, dema = N-methylbis(2-aminoethyl)amine).

3.3.3 Elution Order and Absolute Configuration

The relationship between absolute configurations and the elution order was studied by Fujita et al. [57]. The results for various complexes and P-cellulose(adsorbent)—HCl(eluent) are shown in Table 3.5 [69], in which the sign of the longer wavelength CD peak in the first absorption band region is given for the first eluted fraction. The results show that for the [Co(N)$_6$]-type complexes the enantiomers with the Δ-configuration are eluted first except for the [Co(tn)$_3$]$^{3+}$ complex, and that for the mixed ligand univalent complexes the enantiomers with the Λ-configuration are eluted first.

The results for the DEAE-cellulose—HCl system (Table 3.6 [69]) show that for the complexes listed the enantiomers with the Δ-configuration are eluted first. For the ion-exchange Sephadex—optically active eluent system, it is difficult to find a simple relationship between the elution order and the absolute configuration.

The resolution by the mentioned column chromatography for chiral complexes is based on the interaction between a chiral adsorbent and a chiral eluent. From such a point of view the association constants have been determined for the ion pairs, Δ- and Λ-[Co(en)$_3$]$^{3+}$ with L(+)-tartrate dianion and with L(+)-tartratoantimonate(III) dianion ((+)-tartan), on the basis of CD measurements at two wave-

Table 3.5. Correlation between Sign CD Peak of the First Eluted Enantiomers and Their Absolute Configuration

P-cellulose

Isomer	Sign of the longer wavelength CD peak	Absolute configuration
[Co(en)$_3$]$^{3+}$	−	Δ
[Co(chxn)$_3$]$^{3+}$, lel$_3$	−	Δ
[Co(R-pn)$_3$]$^{3+}$, lel$_3$	−	Δ
[Co(tn)$_3$]$^{3+}$	+	Λ
[Co(bpy)$_3$]$^{3+}$	−^{a}	Δ
cis-[Co(NH$_3$)$_2$(en)$_2$]$^{3+}$	−	Δ
u-fac-[Co(dien)$_2$]$^{3+}$	−	ΔΛΛ
[Co(penten)]$^{3+}$	−	ΔΛΛ
[Co(mepenten)]$^{3+}$	−	ΔΛΛ
[Co(ox)(en)$_2$]$^+$	+	Λ
cis-[Co(NO$_2$)$_2$(en)$_2$]$^+$	+	Λ
cis-α-[Co(NO$_2$)$_2$(trien)]$^+$	+	Λ
u-cis-[Co(ida)(dien)]$^+$	−	ΛΔΛ

^{a} Sign of the main CD peak, not of the longer wavelength peak

Table 3.6. Correlation between Sign of CD Peak of the First Eluted Enantiomers and Their Absolute Configuration

DEAE-Cellulose

Complex	Sign of the longer wavelength CD peak	Absolute configuration
$[Co(ox)_3]^{3-}$	−	Δ
$[Co(rdas)_3]^{3-}$ [a]	−	Δ
$[Co(ox)_2(en)]^-$	−	Δ
cis-$[Co(ida)_2]^-$	−	$\Delta\Lambda\Delta$
$[Co(edta)]^-$	−	$\Delta\Lambda\Delta$

[a] rdas = rac-diaminosuccinate

Table 3.7. Association Constants

25 °C, $\mu = 0.1$

Complex	nm	L(+)-tart^{2-}	L(+)-tartan^{2-}
$\Lambda[Co(en)_3]^{3+}$	430	13.0 ± 0.1	46.6 ± 0.7
	455	13.6 ± 0.2	47.3 ± 0.6
$\Delta[Co(en)_3]^{3+}$	430	10.5 ± 0.4	25.5 ± 0.9
	455	11.7 ± 0.4	26.8 ± 0.6

lengths [70,71] (Table 3.7). In both cases the values for the Λ enantiomers are larger than the corresponding values for the Δ enantiomers; the values for (+)-tartan are considerably larger than those for tart. This indicates that (+)-tartan is an efficient eluent.

Thus, chromatography on Sephadex ion-exchangers is very effectively applied to multivalent complex cations. Jensen and Woldbye [72] have reviewed the optical activity of coordination compounds; resolutions of racemic octahedral transition metal complexes through both diastereoisomer and chromatographic techniques are summarized.

3.3.4 Neutral and Anionic Complexes

Gillard and Spencer [73] isolated anionic and neutral complexes of cobalt(III) with glycyl-L-histidine by combining anion-exchange resin chromatography and Sephadex G-10 chromatography. Later, Gillard et al. [74] used chromatography on a column of Sephadex G-10 for separating complexes of different charges and/or geometries: the column gave completely separated bands of $[Co(en)_3]^{3+}$, [Co(amino-acidate)-(en)_2]^{2+} and [Co(amino-acidate)_3], and also the bands of mer and fac isomers of the neutral complex; this separation of geometrical isomers was more efficient on a Sephadex C-25 column, using water as eluent. When fac-$[Co(ox)(L-asp)(en)]^-$ isomer separated from the corresponding mer isomer on a column of the H^+-form Dowex 50W-X8 was rechromatographed on a column of Sephadex G-15, [36] its diastereoisomers (Δ and Λ) were separately eluted with water.

A complete resolution of *fac*-tris(β-alaninato)cobalt(III), *fac*-[Co(β-ala)$_3$], was achieved by Yoneda and Yoshizawa [75] from a column of CM Sephadex C-25 in Na$^+$ form (3 × 113 cm) with a 0.1 mol/dm^3 sodium (+)-tartrate in 30% aqueous ethanol solution. The isomer exhibiting a (−)-signed dominant CD peak in the first absorption band region was faster eluted. *fac*[Co(β-ala)$_n$(α-am)$_{3-n}$] has been studied by Yoneda and his co-workers [76,77].

For the separation of isomeric L-aspartato-L-asparginato complexes on Sephadex QAE-A25 see 4.3.

3.4 Chromatography using Cellulose Powder, Paper, or Alumina

3.4.1 Mixed Ligand Complexes with Diamines

The best example for cellulose column and paper chromatography are the mixed ligand cobalt(III) complexes with ethylenediamine and R-propylenediamine [78]; an aqueous solution containing en, R-pn, dil. HCl and CoCl$_2$ · 6 H$_2$O in a molar ratio 1:2:1:1 was aerated over activated charcoal, and the product was then precipitated by evaporating the resulting solution and by adding ethanol to the concentrate. A column (7.5 × 30 cm) of cellulose powder (Whatman, std. grade) in H$_2$O-saturated *n*-butanol served to separate the product in a mixture of H$_2$O-saturated *n*-butanol and dry *n*-butanol (5:8), which was then eluted by a 97% BuOH − 2% 10 mol/dm^3 HCl mixture under a reduced pressure. The first three bands contained Δ(−)[Co(R-pn)$_3$]Cl$_3$, Δ(−)[Co(en)(R-pn)$_2$]Cl$_3$ and Δ(−)[Co(en)$_2$(R-pn)]Cl$_3$, respectively, and the fourth band contained all the Λ(+)-complexes and racemic [Co(en)$_3$]Cl$_3$.

Then, each chloride was extracted from each of the former three fractions by shaking repeatedly with water and isolated as solids by evaporating the solution and then by adding ethanol. After repeated chromatography Δ-[Co(R-pn)$_3$]Cl$_3$, Δ-[Co(en)-(R-pn)$_2$]Cl$_3$ · 2 H$_2$O and Δ-[Co(en)$_2$(R-pn)]Cl$_3$ · H$_2$O were obtained as pure compounds.

The fourth fraction was rechromatographed with H$_2$O-saturated *n*-BuOH:dry *n*-BuOH:conc. HClO$_4$ (800:200:13). The initial fractions contained Λ-[Co(R-pn)$_3$]Cl$_3$ and the final portion contained [Co(en)$_3$]Cl$_3$. The combined middle fractions were extracted into water and solid complexes were recovered finally as a mixture of chlorides, *via* triiodides and iodides. The chloride mixture was then subjected to paper chromatography using Whatman Grade 3 paper and developed with *n*-BuOH:H$_2$O:60% HClO$_4$ (70:20:10). The separated four bands corresponded to Λ(+)[Co(R-pn)$_3$]$^{3+}$, Λ(+)[Co(en)(R-pn)$_2$]$^{3+}$, Λ(+)[Co(en)$_2$(R-pn)]$^{3+}$ and [Co(en)$_3$]$^{3+}$ in the order of elution. Finally Λ(+)[Co(en)(R-pn)$_2$](ClO$_4$)$_3$ · H$_2$O and Λ(+)[Co(en)$_2$-(R-pn)](ClO$_4$)$_3$ · H$_2$O were isolated from the second and third bands of the chromatogram.

At the same time, Dwyer et al. determined equilibrium concentrations for the isomers in the reaction mixtures with varied molar ratios of en/pn. For en/pn = 1/2, the paper chromatography on Whatman 3 MM with *sec*-BuOH:H$_2$O:10 mol/dm^3 HCl (70:20:10), the separated bands corresponded to Δ-[Co(R-pn)$_3$]$^{3+}$, Δ-[Co(en)-(R-pn)$_2$]$^{3+}$, Δ-[Co(en)$_2$(R-pn)]$^{3+}$ and Λ-isomers + [Co(en)$_3$]$^{3+}$. The last fraction was rechromatographed on a paper with *n*-BuOH:H$_2$O:60% HClO$_4$ (70:20:10), four bands being completely separated. The experiment of en/pn = 2/1 was carried out in the described manner. In a similar experiment with the reaction mixture without ethylenediamine, the paper chromatogram was developed with n-BuOH:H_2O: 10 mol/dm^3 HCl (60:30:10). Thus, the equilibrium concentrations of the isomers separated by paper chromatography were estimated spectrometrically (see 4.1).

The cellulose column chromatography was soon applied to equilibrium mixtures of the complexes with R,S-propylenediamine [79]; the first fraction contained $\Lambda(+)[\text{Co(S-pn)}_3]^{3+}$ and $\Delta(-)[\text{Co(R-pn)}_3]^{3+}$, the second fraction $\Lambda(+)[\text{Co(S-pn)}_2$-$(\text{R-pn})]^{3+}$ and $\Delta(-)[\text{Co(R-pn)}_2(\text{S-pn})]^{3+}$, and the third fraction $\Lambda(+)[\text{Co(S-pn)}$-$(\text{R-pn})_2]^{3+}$, $\Delta(-)[\text{Co(R-pn)(S-pn)}_2]^{3+}$, $\Lambda(+)(\text{Co(R-pn)}_3]^{3+}$ and $\Delta(-)[\text{Co(S-pn)}_3]^{3+}$, the racemates obtained from the first and second fractions being resolved by the usual method of diastereoisomer formation.

Bang et al. [80] developed a preparative paper chromatographic method to separate all possible isomers of mixed ethylenediamine(en)-trimethylenediamine(tn) complex ions by the ascending method on 1 mm thick Whatman 3 MM paper (46 × 57 cm). The complexes were prepared by aerial oxidation of a mixture of cobalt(II) salt, tn, and en over activated charcoal. The resulting salt mixture (2.5 g) was applied to 16 sheets along the shorter edge. After elution with n-butanol:acetone:90% phenol:pyridine:benzene:water:80% acetic acid (14:14:30:14:14:4:7 v/v) (25 °C, ca. 150 h), the chromatogram consisted of four well separated zones, $[\text{Co(en)}_3]^{3+}$, $[\text{Co(en)}_2(\text{tn})]^{3+}$, $[\text{Co(en)(tn)}_2]^{3+}$, and $[\text{Co(tn)}_3]^{3+}$ from top to bottom. Two zones with the desired complexes were extracted by water to isolate $[\text{Co(en)}_2(\text{tn})]\text{Cl}_3 \cdot 3.5\,H_2O$ and $[\text{Co(en)(tn)}_2]\text{Cl}_3 \cdot 3.5\,H_2O$. Resolution of these complexes was achieved by the diastereoisomer formation to conclude that $(+)_{589}$ enantiomers have the same absolute configuration, Λ, as $(+)_{589}[\text{Co(en)}_3]^{3+}$ and $(-)_{589}[\text{Co(tn)}_3]^{3+}$ from ORD and CD data in the first absorption band region.

3.4.2 Tris(Amino-acidato) Complex

In 1954, Krebs and Rasche [81] reported on the partial resolution of mer-$[\text{Co(gly)}_3]$ on a starch column by elution with water. Later, Douglas and Yamada [82] reported on the partial resolution of fac-$[\text{Co(gly)}_3]$; the fac isomer was dissolved in water containing 10% KCl in order to increase the solubility of the isomer, and the solution was eluted from a potato starch column (2.2 × 70 cm) with 10% KCl solution.

Among the tris(L-alaninato)cobalt(III) isomers, $[\text{Co(L-ala)}_3]$, the fac-Λ isomer is insoluble in water and only soluble in 50% sulfuric acid, mer-Λ is sparingly soluble in water, mer-Δ isomer is very soluble in water and sparingly soluble in ethanol, and fac-Δ is soluble in water and sparingly soluble in ethanol. By these properties, Dunlop and Gillard [83] isolated the four isomers; water-insoluble fac-Λ was first separated from the other isomers by filtration of the reaction product. Two mer diastereoisomers were then obtained by the fractional crystallization from the aqueous solution. The remaining material was extracted with ethanol, and the solution was evaporated to dryness. The residues were dissolved in water, and the aqueous solution was then chromatographed on a column of alumina. The fourth isomer, fac-Δ, was eluted after the still existing small amount of mer-Δ isomer.

Gillard and Payne [84] also used alumina chromatography for the preparations of the isomers of tris(L-valinato) complex, [Co(L-val)₃], tris(L-leucinato) complex, [Co(L-leu)₃], and tris(γ-methyl-L-glutamato) complex, [Co(γ-Me-L-glut)₃]; the complexes were prepared by a new method, that is, by aerial oxidation of aqueous solutions of cobalt(II) amino-acidates with hydrogen peroxide, yielding mainly *mer*-isomers. In the chromatographic separations of the complexes [Co(L-val)₃], [Co(L-leu)₃] and [Co(L-ala)₃] on acid-washed alumina by elution with an aqueous ethanol, the isomers appeared in the order *mer*-Δ, *mer*-Λ, *fac*-Δ, while in the case of the [Co(γ-Me-L-glut)₃] complex on neutral alumina, the *mer*-Λ isomer was eluted faster than the *mer*-Δ isomer. Ćelap et al. [85] reported that *mer* and *fac* isomers of the [Co(β-ala)₃] complex were separated on a column of an acid alumina (Merck for chromatography) by elution with water.

3.4.3 Complexes with β-Keto Enolate or β-Diketonate

For the tris-cobalt(III) complex with the (+)-hydroxymethylenecamphorate ion (+ hmc), [Co(+ hmc)₃], four diastereoisomers, *mer*-Λ, -Δ, *fac*-Λ and -Δ are possible (see Fig. 4.6). Dunlop et al. [86] found two isomers, *mer*-Λ and *fac*-Λ, by chromatography on alumina; the complex was prepared by the reaction of Na₃[Co(CO₃)₃] · 3 H₂O and the ligand (+ hmc) in H₂O/C₆H₆ (1:1) and extracted with ligroin. The solid complex obtained by evaporating the ligroin solution was dissolved in benzene and chromatographed on an alumina column. Elution with benzene separated the *mer*-Λ and *fac*-Λ isomers in this order. Springer Jr. et al. [87] did a study on the stereoisomers of tris-complex with (+)-3-acetyl-camphorate ((+)-atc; see 4.6-I); the complex, [Co{(+)-atc}₃], was prepared from Na₃[Co(CO₃)₃] · 3 H₂O according to the method of Dunlop et al. [86]. By chromatography, all the possible isomers were isolated, the order of elution being *mer*-Δ, *mer*-Λ, *fac*-Λ and *fac*-Δ.

Thin layer chromatography (TLC) with silicagel [88] is useful to separate the diastereoisomers of neutral complexes with β-keto enolate ligands; the tris-complexes of (+)- and (−)-hydroxymethylenecarvone (Hhmcar; see 4.6-II) and (+)-hydroxymethylenepulegone (Hhmpue; see 4.6-III) were prepared by a method similar to that of Gillard et al. [86] Since attempts to separate the diastereoisomers by alumina column chromatography were not successful, precoated preparative TLC plates with 3-mm layer of Silica Gel F-254 were used. The ascending technique was employed with multiple development using a 3:1 (v/v) *n*-pentane/diethylether. The separated bands from the plates were extracted with methanol. For [Co{(−)-hmcar}₃], the four isomers were eluted in the order *fac*-Δ, *mer*-Λ, *mer*-Δ and *fac*-Λ. The superior resolving power of TLC over column chromatography helped to separate [Co{(+)-atc}₃] [89] with *n*-pentane/diethylether (4:1) into four isomers, *mer*-Δ, *mer*-Λ, *fac*-Λ and *fac*-Δ in decreasing order of elution.

Chromatography with D(+)-lactose hydrate columns was used in the partial resolution of tris(acetylacetonato)cobalt(III) complex [90]; partial resolution of [Co(acac)₃], and *mer*- and *fac*-[Co(bzac)₃] (bzac = benzoylacetonate) is also effected by column chromatography on (+)-lactose. The [Co(bzac)₃] complex prepared from Na₃[Co(CO₃)₃] · 3 H₂O was separated into its *mer* and *fac* geometrical isomers by column chromatography on Florisil [92] (Florodin Co., superior to alumina), prior to the resolution. For the resolution, (+)-lactose dried at 110 °C and then sieved to 100 mesh, and a chromatographic column (3.9 × 230 cm), which narrowed to 3.0 cm at the midpoint, were used. The lactose was tightly packed by periodic vibration from an electric vibrator attached to the top of the column. The complex in benzene was eluted with 1:1 (v/v) benzene/hexane; the optical rotation of the eluate was monitored by polarimetry. After this chromatographic procedure, the less soluble racemate was crystallized out from a benzene-hexane solution of the partially resolved complex, whereby the degree of resolution increased about tenfold.

Other complexes such as $[Co(acac)_n(tfac)_{3-n}]$ (tfac = trifluoroacetylacetonate; n = 1, 2, 3)[93], $[Co(acac)_2(L-am)]$ (L-am = L-ala, L-val, N-methyl-L-ala or N-methyl-L-val)[94,95], and $[Co(acac)(L-val)_2]$[94] were separated into geometrical isomers or diastereoisomers by column chromatography on alumina.

Thin-layer chromatography has found widespread use in investigations of metal complexes. Dutta et al.[96] reviewed a number of studies of 1968~1977 with emphasis on TLC and paper chromatography. Ćelap et al.[97] reported on the effect of the composition and structure of cobalt(III) complexes on their R_F values, in which the chromatographic separation, on silica gel, of 75 complexes using several solvent system are investigated.

3.5 References

1. King, E. L., Walters, R. R.: J. Am. Chem. Soc. *74*, 4471 (1952)
2. Mori, M., Shibata, M., Azami, J.: Nippon Kagaku Zasshi *76*, 1003 (1955)
3. Mori, M., Shibata, M., Nanasawa, M.: Bull. Chem. Soc. Jpn. *29*, 947 (1956)
4. Mori, M., Shibata, M., Hori, K.: Bull. Chem. Soc. Jpn. *34*, 1809 (1961)
5. Yamamoto, Y., Nakahara, A., Tsuchida, R.: J. Chem. Soc. Jpn. *75*, 232 (1954)
6. Joergensen, S. M.: Z. Anorg. Chem. *17*, 463 (1898)
7. Joergensen, S. M.: Z. Anorg. Chem. *17*, 477 (1898)
8. Stefanović, G., Janjić, T.: Anal. Chim. Acta *11*, 550 (1954)
9. Stefanović, G., Janjić, T.: Anal. Chim. Acta *19*, 488 (1954)
10. Druding, L. F., Kauffmann, G. B.: Coordin. Chem. Rev. *3*, 409 (1968)
11. Legg, J. I., Cooke, D. W.: Inorg. Chem. *4*, 1576 (1965)
12. Legg, J. I., Cooke, D. W.: Inorg. Chem. *5*, 594 (1966)
13. Legg, J. I., Cooke, D. W.: J. Am. Chem. Soc. *89*, 6854 (1967)
14. Schoenberg, L. N., Cooke, D. W., Liu, C. F.: Inorg. Chem. *8*, 2386 (1968)
15. Hosaka, K., Nishikawa, H., Shibata, M.: Bull. Chem. Soc. Jpn. *42*, 277 (1969)
16. Yamada, S., Hidaka, J., Douglas, B. E.: Inorg. Chem. *10*, 2187 (1971)
17. Hidaka, J., Yamada, S., Douglas, B. E.: J. Coord. Chem. *2*, 123 (1972)
18. Legg, J. I., Neal, J. A.: Inorg. Chem. *12*, 1805 (1973)
19. Froebe, L. R., Yamada, S., Hidaka, J., Douglas, B. E.: J. Coord. Chem. *1*, 183 (1971)
20. Oonishi, I., Shibata, M., Marumo, F., Saito, Y.: Acta Cryst. *B29*, 2448 (1973)
21. Takeuchi, M., Matsuda, T., Shibata, M., Oonishi, I., Saito, Y.: Chem. Lett. *1973*, 531
22. Okamoto, K., Hidaka, J., Shimura, Y.: Bull. Chem. Soc. Jpn. *44*, 1601 (1971)
23. Okamoto, K., Hidaka, J., Shimura, Y.: Bull. Chem. Soc. Jpn. *46*, 473 (1973)
24. Legg, J. I., Cooke, D. W., Douglas, B. E.: Inorg. Chem. *6*, 700 (1967)
25. Halloran, L. J., Legg, J. I.: Inorg. Chem. *13*, 2193 (1974)
26. Jordan, W. T., Legg, J. I.: Inorg. Chem. *13*, 2271 (1974)
27. Matsuoka, N., Hidaka, J., Shimura, Y.: Bull. Chem. Soc. Jpn. *40*, 1868 (1967)
28. Matsuoka, N., Hidaka, J., Shimura, Y.: Inorg. Chem. *9*, 719 (1970)
29. Matsuoka, N., Hidaka, J., Shimura, Y.: Bull. Chem. Soc. Jpn. *45*, 2491 (1972)
30. Matsuoka, N., Hidaka, J., Shimura, Y.: Bull. Chem. Soc. Jpn. *48*, 458 (1975)
31. Kobayashi, K., Shibata, M.: Bull. Chem. Soc. Jpn. *48*, 2561 (1975)
32. Kanazawa, S., Shibata, M.: Bull. Chem. Soc. Jpn. *44*, 2424 (1971)
33. Nakai, K., Kanazawa, S., Shibata, M.: Bull. Chem. Soc. Jpn. *45*, 3544 (1972)

34. Nakashima, S., Shibata, M.: Bull. Chem. Soc. Jpn. *47*, 2069 (1974)
35. Nakashima, S., Shibata, M.: Bull. Chem. Soc. Jpn. *48*, 3128 (1975)
36. Takeuchi, M., Shibata, M.: Bull. Chem. Soc. Jpn. *47*, 2797 (1974)
37. Watabe, M., Onuki, K., Yoshikawa, S.: Bull. Chem. Soc. Jpn. *48*, 687 (1975)
38. Yoneda, H., Sakaguchi, U., Nakashima, Y.: Bull. Chem. Soc. Jpn. *48*, 209 (1975)
39. Watabe, M., Kawaai, S., Yoshikawa, S.: Bull. Chem. Soc. Jpn. *49*, 1845 (1976)
40. Yoshino, Y., Sugiyama, H., Nogaito, S., Kinoshita, H.: Sci. Papers Coll. General Eduction, Univ. Tokyo *16*, 57 (1966)
41. Brubaker, G. R., Legg, J. I., Douglas, B. E.: J. Am. Chem. Soc. *88*, 3446 (1966)
42. Yoshikawa, Y., Yamasaki, K.: Inorg. Nucl. Chem. Lett. *4*, 697 (1968)
43. Yoshikawa, Y., Yamasaki, K.: Inorg. Nucl. Chem. Lett. *6*, 523 (1970)
44. Yoshikawa, Y., Yamasaki, K.: Coord. Chem. Rev. *28*, 205 (1979)
45. Dwyer, F. P., Sargeson, A. M., James, L. B.: J. Am. Chem. Soc. *86*, 590 (1964)
46. MacDermott, T. E.: Inorg. Chim. Acta *2*, 81 (1968)
47. Kojima, M., Yoshikawa, Y., Yamasaki, K.: Inorg. Nucl. Chem. Lett. *9*, 689 (1973)
48. Harnung, S. E., Kallesøe, S., Sargeson, A. M., Schaeffer, C. E.: Acta Chem. Scand. *A28*, 365 (1974)
49. Toftlund, H., Pedersen, E.: Acta Chem. Scand. *26*, 4019 (1972)
50. Tanimura, T., Ito, H., Fujita, J., Saito, K., Hirai, S., Yamasaki, K.: J. Coord. Chem. *3*, 161 (1973)
51. Kojima, M., Yoshikawa, Y., Yamasaki, K.: Bull. Chem. Soc. Jpn. *46*, 1687 (1973)
52. Fujita, J., Ogino, H.: Chem. Lett. *1973*, 57
53. Kojima, M., Fujita, J.: Bull. Chem. Soc. Jpn. *50*, 3237 (1977)
54. Kojima, M., Yamada, H., Ogino, H., Fujita, J.: Bull. Chem. Soc. Jpn. *50*, 2325 (1977)
55. Fujita, M., Yoshikawa, Y., Yamatera, H.: Chem. Lett. *1974*, 1515
56. Fujita, M., Yoshikawa, Y., Yamateray, H.: Chem. Lett. *1975*, 473
57. Fujita, M., Yoshikawa, Y., Yamatera, H.: Chem. Commun. *1975*, 941
58. Keene, F. R., Searle, G. H., Yoshikawa, Y., Imai, A., Yamasaki, K.: Chem. Commun. *1970*, 784
59. Yoshikawa, Y., Yamasaki, K.: Bull. Chem. Soc. Jpn. *45*, 179 (1972)
60. Keene, F. R., Searle, G. H.: Inorg. Chem. *11*, 148 (1972)
61. Keene, F. R., Searle, G. H.: Inorg. Chem. *13*, 2173 (1974)
62. Mason, S. F., Norman, B. J.: J. Chem. Soc. A. *1966*, 307
63. Searle, G. H., Larsen, E.: Acta Chem. Scand. *A30*, 143 (1976)
64. Kojima, M., Iwagaki, M., Yoshikawa, Y., Fujita, J.: Bull. Chem. Soc. Jpn. *50*, 3216 (1977)
65. Yoshikawa, Y., Yamasaki, K.: Bull. Chem. Soc. Jpn. *46*, 3448 (1973)
66. Yoshikawa, Y., Fujii, E., Yamasaki, K.: Bull. Chem. Soc. Jpn. *45*, 3451 (1972)
67. Yoshikawa, Y., Kondo, A., Yamasaki, Y.: Inorg. Nucl. Chem. Lett. *12*, 351 (1976)
68. Kojima, M., Iwagaki, M., Yoshikawa, Y., Fujita, J.: Bull. Chem. Soc. Jpn. *50*, 3216 (1977)
69. Yamasaki, K., Yoshikawa, Y.: Revue Roumaine Chimic *22*, 801 (1977)
70. Yoneda, H., Taura, T.: Chem. Lett. *1977*, 63
71. Taura, T., Nakazawa, H., Yoneda, H.: Inorg. Nucl. Chem. Lett. *13*, 603 (1977)
72. Jensen, H. P., Woldbye, F.: Coord. Chem. Rev. *29*, 213 (1979)
73. Gillard, R. D., Spencer, A.: J. C. S. Dalton *1972*, 902
74. Gillard, R. D., Laurie, S. H., Price, D. C., Phipps, D. A., Weick, C. F.: J. C. S. Dalton *1974*, 1385
75. Yoneda, H., Yoshizawa, T.: Chem. Lett. *1976*, 707
76. Yamazaki, S., Yukimoto, T., Yoneda, H.: J. Chromatogr. *175*, 317 (1979)
77. Yukimoto, T., Yoneda, H.: J. Chromatogr. *210*, 477 (1981)
78. Dwyer, F. P., MacDemott, T. E., Sargeson, A. M.: J. Am. Chem. Soc. *85*, 2913 (1963)
79. Dwyer, F. P., Sargeson, A. M., James, L. B.: J. Am. Chem. Soc. *86*, 590 (1964)
80. Bang, O., Engberg, A., Rasmussen, K., Woldbye, F.: Acta Chem. Scand. *A29*, 749 (1975)
81. Krebs, H., Rasche, R.: Z. Anorg. Allg. Chem. *276*, 273 (1954)
82. Douglas, B. E., Yamada, S.: Inorg. Chem. *4*, 1561 (1965)
83. Dunlop, J. H., Gillard, R. D.: J. Chem. Soc. *1965*, 6531
84. Gillard, R. D., Payne, N. C.: J. Chem. Soc. A. *1969*, 1197
85. Celap, M. B., Niketić, S. R., Janjic, T. J., Nikolić, V. N.: Inorg. Chem. *6*, 2063 (1967)
86. Dunlop, J. H., Gillard, R. D., Ugo, R.: J. Chem. Soc. A. *1966*, 1540

87. Springer, C. S., Jr., Sievers, R. E., Freibush, B.: Inorg. Chem. *10*, 1242 (1971)
88. Everett, G. W., Jr., Chen, Y. I.: J. Am. Chem. Soc. *92*, 508 (1970)
89. King, R. M., Everett, G. W., Jr.: Inorg. Chem. *10*, 1237 (1971)
90. For example, Moeller, T., Gulyas, E.: J. Inorg. Nucl. Chem. *5*, 245 (1958): Collman, J. P., Blair, R. P., Marshall, R. L., Slade, L.: Inorg. Chem. *2*, 576 (1963)
91. Fay, R. C., Girgis, A. Y., Klabunde, U.: J. Am. Chem. Soc. *92*, 7056 (1970)
92. Fay, R. C., Piper, T. S.: J. Am. Chem. Soc. *84*, 2303 (1962)
93. Fay, R. C., Piper, T. S.: J. Am. Chem. Soc. *85*, 500 (1963)
94. Seematter, D. J., Brushmiller, J. G.: J.C.S. Chem. Comm. *1972*, 2177
95. Everett, G. W., Jr., Finney, K. S., Brushmiller, J. G., Seematter, D. J., Wingert, L. A.: Inorg. Chem. *13*, 536 (1974)
96. Dutta, R. L., Ray, R. K., Kauffman, G. E.: Coord. Chem. Rev. *28*, 23 (1979)
97. Celap, M. B., Vučković, G., Malinar, M. J., Janjić, T. J., Radivojša, P. N.: J. Chromatogr. *196*, 59 (1980)

4 Stereoselectivity in Complexes with Less Puckered Chelate Rings

4.1 Historical Background

When ethylenediamine molecules in $[Co(en)_3]^{3+}$ are replaced by an optically active diamine such as R(—)-propylenediamine (R-pn), the resulting two optical isomers or diastereoisomers are no longer in equal amounts. This preference for one optical isomer or one diastereoisomer over the other has been called "ligand stereospecificity"[1]. In the same sense, the term "stereoselectivity" has been used by Dunlop and Gillard, who defined it as "the behavior of molecular diastereoisomers"[2]. In this chapter, the word "stereoselectivity" is used to mean inclination in abundance of the diastereoisomers of one geometrical isomer. This word differs from either "stereoselective reaction" or "stereospecific reactions".[15]

Tris(propylenediamine)cobalt(III) salts were first synthesized from racemic propylenediamine[3]. Later, the synthesis[7] with the optically active base showed that the action of S(+)-pn on $[CoCl_2(R-pn)_2]^+$ gave a mixture of $(—)[Co(R-pn)_3]^{3+}$ and $(+)[Co(S-pn)_3]^{3+}$ rather than $[Co(R-pn)_2(S-pn)]^{3+}$. After all the salts proved to be $(+)[Co(S-pn)_3]Br_3 \cdot 2 H_2O$ from the S-base and $(—)[Co(R-pn)_3]Br_3 \cdot 2 H_2O$ from the R-base.[5]

However, oxidizing an aqueous solution of Co(II) salt and racemic 1,2-*trans*-cyclopentanediamine (cptn) resulted in *trans*-$[CoCl_2\{(+)-cptn\}_2]^+$ and *trans*-$[CoCl_2\{(—)-cptn\}_2]^+$, without *trans*-$[CoCl_2\{(+)-cptn\}\{(—)-cptn\}]^+$. When a solution of *trans*-$[CoCl_2\{(+)-cptn\}_2]^+$ was heated, it was converted to *cis* and $(—)[CoCl_2\{(+)-cptn\}_2]^+$ was preferentially formed. The action of $(+)$-cptn on the $(—)[CoCl_2\{(+)-cptn\}_2]^+$ complex resulted in $(—)[Co\{(+)-cptn\}_3]^{3+}$, whereas the action of $(—)$-cptn resulted in a 2:1 mixture of $(—)[Co\{(+)-cptn\}_3]^{3+}$ and $(+)[Co\{(—)-cptn\}_3]^{3+}$. Through these and the other studies, early workers thought, in general, that an optically active ligand favored the formation of one isomer to the exclusion of the other.

In 1959 Corey and Bailar[7] dealt with stereo specific effects in complex ions. The results of the conformational analysis for the $[Co(en)_3]^{3+}$ ion showed that one conformational form of a chelated en, "*lel*", is more stable than the other, "*ob*", by *ca.* 0.6 kcal/mol. For the tris(propylenediamine)cobalt(III) ion, the three most stable pn chelate rings are suggested to have equatorial conformation with

---

[15] The reactions by which stereoisomers are formed or used up at different rates are called stereoselective reactions, while the reactions by which specific stereochemical compounds are converted to specific stereochemical products are called stereospecific reactions.

respect to methyl groups. The energy difference between this equatorial and the axial conformation was evaluated to be more than 2 kcal/mol.

In the same year, Dwyer et al. [8] isolated $\Delta(-)$ and $\Lambda(+)[Co(R\text{-}pn)_3]I_3$. In order to determine the degree of ligand stereospecificity, they equilibrated both diastereoisomers by shaking with water/charcoal. The Δ-RRR $\rightleftarrows \Lambda$-RRR equilibrium contains *ca.* 14.8% Λ-RRR, just as after the synthesis from the base with charcoal. Furthermore, Dwyer et al. [9] obtained all possible isomers of $[Co(en)_{3-n}\text{-}(R\text{-}pn)_n]^{3+}$ (n = 1, 2, 3) from an aerated reaction mixture of components with charcoal by a combination of cellulose column chromatography and paper chromatography. From the equilibrium concentrations for the isomers (see 3.4) the free energy differences between each pair of diastereoisomers were evaluated (Table 4.1). For each R-pn introduced, the free energy difference increases by *ca.* 0.5 kcal, close to the calculated energy difference 0.6 kcal/mol. Subsequently, Dwyer et al. [10] isolated the optical isomer, $\Lambda(+)[Co(S\text{-}pn)_2(R\text{-}pn)]I_3$ and $\Delta(-)[Co(R\text{-}pn)_2(S\text{-}pn)]I_3$ (see 3.4).

Thus these works stimulated great interest in the stereoselective formation of diastereoisomers.

Table 4.1. Formation Ratios for $[Co(en)_{3-n}(R\text{-}pn)_n]^{3+}$

n	$\Delta(-)/\Lambda(+)$	ΔG_{obs} kcal mol^{-1}, 25 °C	ΔG_{calc} kcal mol^{-1}, 25 °C
1	2.1/1	0.45	0.6
2	7.5/1	1.2	1.2
3	14.5/1	1.6	1.8

4.2 Complexes with Amino Acids

Figure 4.1 shows the spatial distribution of the alkyl substituents, R, in the four isomers of a tris-(L-α-amino-acidato) complex $[Co(L\text{-}NH_2CHRCOO)_3]$. [11] The substituents are arranged either in a pseudo-axial or in a pseudo-equatorial orientation to the threefold or pseudo-threefold axis of the isomers. The Λ isomers of both *fac* and *mer* have pseudo-equatorial substituents, while the Δ isomers have pseudo-axial ones. In the *mer*-Δ isomer, two substituents are axially disposed at one end of the pseudo-threefold axis and the third at the opposite end. In the *mer*-Λ isomer, there is an exceptionally small distance between two substituents.

Such a steric distribution of the alkyl substituents raises the question whether stereoselectivity would be found in the tris complex with an L-amino acid forming chelate rings much less puckered than the 1,2-diamine chelate rings. The stereoselectivity in tris(L- or D-amino-acidato)cobalt(III) has been mainly studied for the elucidation of the steric interactions between the substituents on the α-carbon atom of the ligand.

There are two ways to estimate the degree of stereoselectivity; in one the products are separated into diastereoisomeric pairs and their relative proportions are determined from their chiroptical properties. Such separation into diastereoisomers is, in

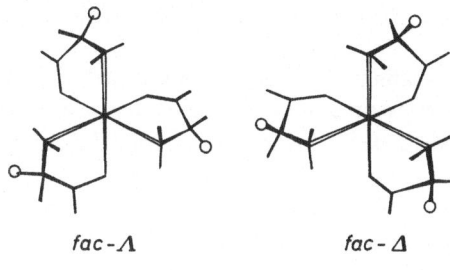

fac-Λ   fac-Δ

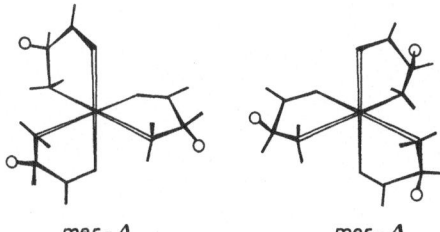

mer-Λ   mer-Δ

Fig. 4.1. Steric distribution of alkyl groups in isomers of [Co(L-am)₃] complexes (from Ref. [11]).

general, attained by chromatography (see the previous chapter). The other way is to obtain an equilibrium mixture by isomerizing a pure isomer and to determine isomer proportions. This way, however, is often unavailable because of the insolubility of one diastereoisomer.

[Co(L-am)₃]: Among the isomers of tris(L-alaninato)cobalt(III) the Λ(+)-*fac* isomer [12] is formed stereoselectively; when the Λ(+)-*mer* or Δ(−)-*mer* isomer was boiled with activated charcoal in water, the Λ(+)-*fac* isomer quickly precipitated. This enormous preponderance of Λ-*fac* over Δ-*fac* was considered to be due to the extremely small solubility of the former displacing the equilibrium completely.

Concerning tris(L-leucinato) complex, Denning and Piper [11] refluxed an *n*-butanol solution of Δ(−)-*mer* isomer until no further change in the CD spectrum was observed. A very small amount of red insoluble *fac* isomer was removed by filtration, the solution was evaporated, and the residue was chromatographed in 85 % ethanol—water on an alumina column. Two determinations of the equilibrium constant were made by CD measurements, and the following equilibrium constant was obtained.

$$K = \frac{[\Delta(-)\text{-isomer}]}{[\Lambda(+)\text{-isomer}]} = 0.742 \pm 0.05 \text{ at } 117 \text{ °C}$$

Consequently the Λ(+)-*mer* isomer is slightly more stable in *n*-butanol solution than the Δ(−)-*mer* isomer. This stability conferred by the pseudo-equatorial nature of the alkyl groups is sufficient to outweigh the steric "clash" interaction.

Gillard and Payne [13] also studied stereoselectivity in tris(L-leucinato) and tris(L-valinato) complexes. Percentage compositions of four preparations are given in Table 4.2. From the yields of the preparative reactions, the formation of Λ(+)-*mer*[Co(L-leu)₃] appears to be slightly more favourable than the formation of

$\Delta(-)$-*mer* isomer. The stabilities of the *fac*-isomers seem to be approximately equal.

Both the $\Lambda(+)$-*mer* and $\Delta(-)$-*mer* tris(valinato) isomers were inverted in boiling ethanol by activated charcoal. The $\Delta(-)$-*mer* isomer was rearranged to give a mixture containing *ca.* 60% $\Delta(-)$-*mer* and 40% $\Lambda(+)$-*mer*. The *fac* isomers could be obtained only by the oxidation with hydrogen peroxide (method D), the yield for each *fac* isomer (*ca.* 15 mg) was considerably less than the amount (*ca.* 0.5 g) for each *mer* isomer. On account of this poor yield, no analysis was made of the percentage compositions of the *fac* isomers.

Table 4.2. Formation Ratios[a] for Isomers found in Various Preparations[b]

	L-Leucine:	Method	Λ/Δ	L-Valine:	Method	Λ/Δ
mer		A	60/40 62/38			
		B	59/41		B	50/50 42/58
		C	74/26			
		D	67/33 68/32		D	51/49 45/55 42/58
fac		B	57/43 61/39		D	—

a The estimated error in each value is *ca.* 5%.
b **A**, from $[Co(NH_3)_6]Cl_3$; **B**, from $Na_3[Co(CO_3)_3] \cdot 3\ H_2O$; **C**, air oxidation of cobalt(II) aminoacidate; **D**, addition of cobalt(II) nitrate and H_2O_2 to sodium aminoacidate.

Denning and Piper [11] worked with *tris*(L-prolinato)cobalt(III). From the molecular models, they considered the alkyl ring of a chelated L-prolinate to have a rigid conformation and the steric interaction in the *mer*-Λ configuration to be prohibitive (Fig. 4.2). The Λ-*fac* configuration shows no steric interaction, while the steric interaction in the Δ-*fac* configuration is slightly greater than that in the Δ-*mer* isomer. In accord with this consideration, the yields in a typical preparation were

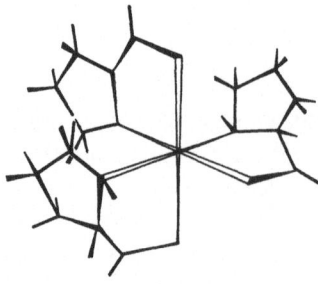

mer - Λ

Fig. 4.2. Steric interaction between two pyrrolidine rings (from Ref. [11])

0.76 g for $\Delta(-)$-*mer*, 0.11 g for $\Delta(-)$-*fac* and 1.48 g for $\Lambda(+)$-*fac* isomer. No $\Lambda(+)$-*mer* isomer was recovered.

The $\Delta(-)$-*fac* isomer isomerized in neutral aqueous solution at 80 °C to an equilibrium mixture with the corresponding $\Delta(-)$-*mer* isomer with the equilibrium constant

$$K = \frac{[\Delta(-)\text{-}mer]}{[\Delta(-)\text{-}fac]} = 3.9 \pm 0.6 \quad \text{at } 80 \text{ °C}$$

Table 4.3. Absorption Spectral and CD Spectral Data on [Co(L-am)$_3$] Complexes[a] (I)

(Band I region, nm)

		λ_{max}	ε	Solvent	λ_{CD}	$\Delta\varepsilon$
L-ala	Λ-*mer*	542	93	H$_2$O	526	+3.3
	Δ-*mer*	544	106	H$_2$O	532	−2.6
	Λ-*fac*	518	188	50% H$_2$SO$_4$	—	
	Δ-*fac*	515	212	H$_2$O	530	−2.9
L-val	Λ-*mer*	532	100	EtOH	527	+2.1
	Δ-*mer*	533	117	EtOH	523	−3.25
	Λ-*fac*	522	151	50% H$_2$SO$_4$	540	+1.1
	Δ-*fac*	532	153	H$_2$O	576, 478	−0.72, −0.97
L-leu	Λ-*mer*	535	104	EtOH	528	+3.2
	Δ-*mer*	537	109	EtOH	529	−2.8
	Λ-*fac*	520	205	50% H$_2$SO$_4$	548	+1.1
	Δ-*fac*	516	207	50% H$_2$SO$_4$	528	−2.6
L-pro	Λ-*mer*	—			—	
	Δ-*mer*	546	103	EtOH	571	−1.9
	Λ-*fac*	541	156	95% H$_2$SO$_4$	565	+4.1
	Δ-*fac*	526	111	H$_2$O	562	−2.6
L-lys	Λ-*mer*	525	100	H$_2$O	532	+1.64
	Δ-*mer*	525	121	H$_2$O	526	−1.82
	Λ-*fac*	—			—	
	Δ-*fac*	—			—	
L-asp	Λ-*mer*	522	126	60% HClO$_4$	517	+2.7
	Δ-*mer*	522	126	60% HClO$_4$	509	−2.3
	Λ-*fac*	520	219	60% HClO$_4$	536	+1.5
	Δ-*fac*	520	219	60% HClO$_4$	526	−1.5
L-glu	Λ-*mer*	532	107	60% HClO$_4$	536	+3.8
	Δ-*mer*	—			—	
	Λ-*fac*	521	186	60% HClO$_4$	545	+1.5
	Δ-*fac*	519	185	60% HClO$_4$	536	−2.3
L-aspn	Λ-*mer*	540	106	60% HClO$_4$	532	+3.8
	Δ-*mer*	543	110	dil alkali soln.	537	−3.1
	Λ-*fac*	—			—	
	Δ-*fac*	518	204	dil. alkali soln.	523	−2.9

a Cited mainly from Ref. [14]

This value shows that the $\Delta(-)$-*mer* isomer is slightly more stable than expected from statistical arguments.

Gillard et al. [14] described stereoselective formation of *mer*-isomers (mainly *mer*-Δ) of [Co(L-lys)$_3$] either in aeration of an aqueous cobalt(II) nitrate—L-lysine solution or in the reaction between [Co(NH$_3$)$_6$]Cl$_3$ and lysine with activated charcoal; *fac*-isomers are only formed in traces. This is peculiar because from [Co(NH$_3$)$_6$]Cl$_3$ usually *fac*-isomers are formed (see 2.5). Two diastereoisomers, *mer*-Δ-[Co(L-lys)$_3$] · 3 H$_2$O and *mer*-Λ-[Co(L-lys)$_3$] · H$_2$O have been isolated by alumina chromatography.

The absorption and CD spectral data for several tris(L-amino-acidato) complexes are cited in Table 4.3.

4.3 Optically Active Aspartate Containing Complexes

[Co(gly)$_{3-n}$(L-asp)$_n$]$^{n-}$: Stereoselective formations of diastereoisomers from L-aspartato-glycinato complexes, [Co(gly)$_{3-n}$(L-asp)$_n$]$^{n-}$ (n = 1, 2, 3), were investigated by Kawasaki et al. [15]; isomer compositions of *mer* and *fac* for Δ and Λ were determined in an aqueous reaction mixture of [CoCO$_3$(gly)$_2$]$^-$, L-aspartate, and activated charcoal at 40~50 °C. The results are cited in Table 4.4. Marked stereoselectivity (more than 90%) is found in every *fac*-Δ isomer, appreciable selectivity (*ca.* 70%) being found in every *mer*-Λ isomer.

Table 4.4. Percentage Compositions of Diastereoisomers in [Co(gly)$_{3-n}$(L-asp)$_n$]$^{n-}$

[Co(gly)$_2$(L-asp)]$^-$		[Co(gly)(L-asp)$_2$]$^{2-}$		[Co(L-asp)$_3$]$^{3-}$	
mer-Λ/—Δ	*fac*-Δ/—Λ	*mer*-Λ/—Δ	*fac*-Δ/—Λ	*mer*-Λ/—Δ	*fac*-Δ/—Λ
61/39	89/11	73/27	95/5	77/23	95/5

Similar experiments with L-glutamato-glycinato complexes, [Co(gly)$_{3-n}$(L-glu)$_n$]$^{n-}$ (n = 1, 2, 3) [16], showed the Δ/Λ ratios in the *fac* isomers to be *ca.* 60/40, 68/32 and 75/25 for the bis(glycinato), bis(L-glutamato) and tris(L-glutamato) complexes, respectively; but no stereoselectivity was found in the *mer* isomers. Further examinations with the [Co(L- or D-ala)(L-asp)$_2$]$^{2-}$ resulted in the Δ/Λ ratio of *ca.* 95/5 in the *fac* isomers and *ca.* 25/75 in the *mer* isomers, irrespective of the used alanine [17].

[Co(L-pro)$_{3-n}$(L- or D-asp)$_n$]$^{n-}$: Na$_3$[Co(CO$_3$)$_3$] · 3 H$_2$O and L-proline were allowed to react [18] in water at 60 °C, and the resulting solution was then allowed to react with L-(or D-)aspartic acid and activated charcoal at 60 °C. The stereoisomers formed were separated by ion-exchange chromatography and the isomeric ratios were estimated spectrophotometrically (Table 4.5). In [Co(L-pro)(L-asp)$_2$]$^{2-}$ (n = 2) the stereoselective formation is found in *fac*-Δ and *mer*-Λ isomers, and in [Co(L-pro)(D-asp)$_2$]$^{2-}$ (n = 2) *fac*-Λ and *mer*-Δ isomers are formed preferentially. In [Co(L-pro)$_2$(L- or D-asp)]$^-$ (n = 1), remarkable differences in stereoselectivity occur depending upon the optical form of the aspartate ion. Both the *fac* and *mer* forms of the complex containing an L-aspartate ligand prefers to form Δ isomers. On the other hand, the *mer* form of the complex containing a D-aspartate prefers to form the Δ isomer, but the *fac* form shows no preference between Λ and Δ isomers.

Table 4.5. Formation Ratios for $[Co(L\text{-pro})_{3-n}(L\text{-asp})_n]^{n-}$ (a) and $[Co(L\text{-pro})_{3-n}(D\text{-asp})_n]^{n-}$ (b)

	n	Isomer[a]	Ratio	Stereoselectivity (%)
(a)	1	*fac*-Δ	4	$\Delta = \sim 100$
		mer-Δ	14	$\Delta = \sim 100$
	2	mer_1-Δ	1	
		mer_2-Δ	1	$\Delta/\Lambda = 20/80$
		mer-Λ	8	
		fac-Δ	5	$\Delta = \sim 100$
(b)	1	*mer*-Λ	1	
		mer_1-Δ	8	
		mer_2-Δ	52	$\Lambda/\Delta = 1/99$
		mer_3-Δ	8	
		fac-Δ	1	
		fac-Λ	1	$\Delta/\Lambda = 50/50$
	2	mer_1-Δ	3	
		mer_2-Δ	4	$\Delta/\Lambda = 95/5$
		mer_3-Δ	28	
		mer-Λ	2	
		fac-Λ	8	$\Lambda = \sim 100$

a mer_1, mer_2, and mer_3 represent three isomers eluted in this order.

The interactions between the side-chain, $R = CH_2COO^-$, of an aspartate ion $(NH_2CHRCOO^-)$ and the adjacent ligand can be classified into four types from the stereo models. (**A**) a favourable interaction through a hydrogen bond between the side-chain —COO$^-$ group and the NH_2 group occupying the apical position (a) in Fig. 4.3; (**B**) an electrostatic repulsive interaction between the β-COO$^-$ group and the ligating O atom occupying the apical position (a); (**C**) interactions of the side-chain of an aspartate ligand with the pyrrolidine ring of the proline ligand, (hydrogen-bonding between the R side-chain of the L-aspartate ligand and the NH group of the L-prolinate ligand is not favorable); (**D**) a marked steric crowding

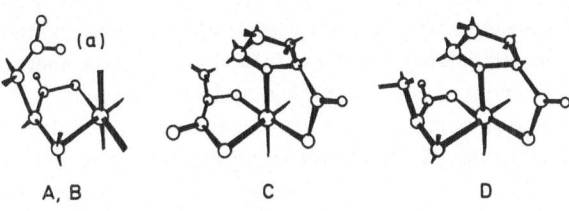

A, B   C   D

Fig. 4.3. Interaction of coordinated aspartate (from Ref. [18])

between the pyrrolidine ring and the R side-chain of the chelated aspartate ion.

These four types are drawn in Fig. 4.3. The types of interaction existing in each isomer are given in Table 4.6, in which the symbol X expresses the steric hindrance between the two pyrrolidine rings. The preferential formation of the Δ isomer in *fac*-[Co(L-pro)(L-asp)$_2$]$^{2-}$ and of the Λ isomer in *fac*-[Co(L-pro)(D-asp)$_2$]$^{2-}$ are explained by the favourable interaction of Type **A**. The preference of the Δ isomers in *mer*-[Co(L-pro)$_2$(D-asp)]$^-$ is explained in the same way. The preferential formation of the Δ isomers in *mer*-[Co(L-pro)$_2$(L-asp)]$^-$ is explained with the unfavorable interaction of Type **D** or **X** in each Λ isomer. Moreover, the poorer yield of the *fac* isomers of [Co(L-pro)$_2$(D-asp)]$^-$ than of the *mer* ones is interpreted in terms of the lack of favourable interaction, Type **A**.

Table 4.6. Interaction of the Side-Chain of Coordinated Aspartate

Complex	Isomer				
	Config.	*fac*	*mer*		
			cis(N)*cis*(O)	*cis*(N)*trans*(O)	*trans*(N)*cis*(O)
Co(L-pro)$_2$(L-asp)$^-$	Δ	A	B	B	A
	Λ	B	D	B(X)	D
Co(L-pro)(L-asp)$_2^{2-}$	Δ	AA	AB	AB	BB
	Λ	BB	BD	AA	BD
Co(L-pro)$_2$(D-asp)$^-$	Δ	B	A	B	A
	Λ	C	B	B(X)	C
Co(L-pro)(D-asp)$_2^{2-}$	Δ	BB	AB	AA	AB
	Λ	AC	AB	BC	BB

A: polar interaction with the amino group in the adjacent ligand.
B: polar interaction with the carboxylate group in the adjacent ligand.
C: some interaction with the pyrrolidine ring.
D: steric interaction with the pyrrolidine ring.
X: steric interaction between two pyrrolidine rings.

[Co(L-aspn)$_{3-n}$(L-asp)$_n$]$^{n-}$: "The green solution" of tricarbonatocobaltate(III) was allowed to react with L-asparagine (L-Haspn) [19] and the resulting solution was allowed to react with L-aspartic acid and activated charcoal at *ca.* 50 °C. The isomers were separated on a column of Sephadex QAE-A25 in Cl$^-$ form. The formation ratios through all the isomers of [Co(L-aspn)$_{3-n}$(L-asp)$_n$]$^{n-}$ are cited in Table 4.7. The marked stereoselectivity is found in both the asparaginato complex (n = 2) and the bis(asparaginato) complex (n = 1); the Δ isomer is preferentially formed in the *fac* isomers of either complex, and the Λ isomer in the *mer* ones. In the tris(L-asparaginato) complex (n = 0), stereoselectivity is found markedly in *fac*-Δ and moderately in *mer*-Λ, this trend being identical with that in [Co(L-asp)$_3$]$^{3-}$.

Table 4.7. Formation Ratios and Stereoselectivities in $[Co(L\text{-aspn})_{3-n}\text{-}$
$(L\text{-asp})_n]^{n-}$ $(n = 0, 1, 2)$

n	Elution order	Isomer	Ratio	Stereoselectivity	
0	L-9	*mer*-Δ	1.5	*mer*-Λ	66%
	L-10	*mer*-Λ	2.9		
	L-11	*fac*-Λ	0.2	*fac*-Δ	95%
	L-12	*fac*-Δ	3.5		
1	L-1	*mer*-Λ	11.5	*mer*-Λ	82%
	L-1	*mer*-Δ	2.5		
	L-2	*fac*-Δ	16.3	*fac*-Δ	~95%
2	L-5	*mer*-Δ	0.4		
	L-6	*mer*-Λ	13.0	*mer*-Λ	83%
	L-6′	*mer*-Δ	3.3		
	L-7	*mer*-Λ	5.1		
	L-8	*fac*-Δ	39.8	*fac*-Δ	~100%

[Co(ox)(L-am)(en)]: A complex of this type, which has the same chromophore as [Co(L-am)₃], gives only four stereoisomers (Fig. 4.4). On this account Shibata and his co-workers investigated the effect of the chelated L-am upon the formation of the stereoisomers following the reaction:

$$[Co(ox)_2(en)]^- + L\text{-}NH_2CHRCOO^- \xrightarrow[\substack{\text{activated charcoal}}]{\text{pH } ca.\ 9.5,\ 40\ °C}$$

$$[Co(ox)(L\text{-am})(en)] \xrightarrow[\substack{\text{chromatographic}\\\text{separation}}]{} \text{four isomers}$$

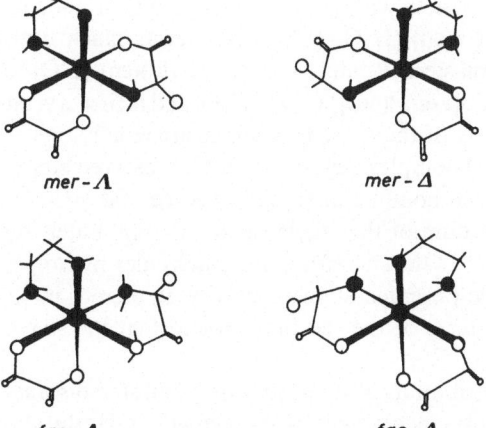

mer-Λ   *mer*-Δ

fac-Λ   *fac*-Δ

Fig. 4.4. Possible four isomers of [Co(ox)-(L-am)(en)]-type complexes (from Ref. [21])

77

The amino acids used were L-alanine [20], L-valine [20], L-isoleucine (Hileu) [20], L-threonine (Hthr) [20], L-methionine (Hmet) [20], L-serine (Hser) [20], L-aspartic acid [21], L-glutamic acid [21], L-leucine [21], L-asparagine [19] and glycine [21]. Table 4.8 gives percent compositions between the diastereoisomers and also between *mer* and *fac* geometrical isomers. High selectivity of the percent compositions around 70% are shown by the *mer*-serinato and *mer*-threoninato complexes as well as the *fac*-isoleucinato, *fac*-valinato and *fac*-asparaginato complexes. More significant stereoselectivity of over 80% is observed in both *mer*- and *fac*-aspartato complexes. These results show that significant interactions of a polar side-chain with other points on the molecular framework are revealed only in the L-aspartate containing complex. Table 4.8 also shows that the *mer*($\Lambda + \Delta$) isomers existing in the system are more abundant than the *fac*($\Lambda + \Delta$) isomers, and that the abundances in the *mer* isomers are roughly 80%, irrespective of L-am ligand.

Table 4.8. Formation Ratios for [Co(ox)(L-am)(en)]

L-am	*mer*-Δ/-Λ	*fac*-Δ/-Λ	*mer*/*fac*
ala	54/46	59/41	81/19
val	46/54	73/27	81/19
ileu	41/59	71/29	82/18
thr	71/29	39/61	84/16
met	54/46	47/53	79/21
ser	68/32	37/63	80/20
asp	87/23	18/82	83/17
glu	60/40	45/55	80/20
leu	58/42	50/50	84/16
asn	77/23	26/74	81/19
gly	50/50	50/50	84/16

4.4 Mixed Diamine-Amino Acid Complexes

[Co-(L-glu)(en)$_2$]$^+$: Dunlop et al. [22,23] observed a kinetically stereoselective formation of this complex; when L-glutamic acid reacts with [Co(CO$_3$)(en)$_2$]ClO$_4$ in aqueous solution, the (+)-enantiomer of the [Co(en)$_2$(H$_2$O)$_2$]$^{3+}$ species reacts much more rapidly than the (−)-complex. The diastereoisomer, (+)-[Co(L-glu)(en)$_2$]ClO$_4$ is much less soluble in water than is (−)-[Co(L-glu)(en)$_2$] ClO$_4$ · H$_2$O, and is first crystallized out from the mixture. The amounts of the diastereoisomers are equal.

A preliminary X-ray study of the (+)-complex revealed that the observed stereoselectivity seemed to result from hydrogen-bonding interaction between the γ-COO$^-$ of the glutamate ligand and a nitrogen atom of the ethylenediamine ring. Later, the details of the X-ray study made it clear [24] "that no strong intramolecular hydrogen-bonding, such as might account for the observed stereoselectivity, is present in the solid". More detailed studies on the course of the reaction showed that the stereoselectivity observed was kinetic in origin [25].

Legg and Steele [26] examined the distribution of Λ(+) and Δ(−) diastereoisomers of the [Co(L-glu)(en)$_2$]ClO$_4$; the reaction between [Co(CO$_3$)(en)$_2$]$^+$, L-H$_2$glu, and activated charcoal at 70 °C produced isomers that were separated by ion-exchange chromatography. They obtained 70% $\Lambda(+)$ and 30% $\Delta(-)$, but found it "difficult to pinpoint any particular steric factor as being dominant, with such a small stereoselectivity".

Buckingham et al. [27] repeated both experiments [25, 26] and demonstrated the lack of kinetic and thermodynamic preference in the reaction systems. They observed low optical activity of both the solution and isolated $[Co(L-glu)(en)_2]^+$ fractions during the course of the reaction [25], indicating the absence of kinetic selectivity. Furthermore, they found that a similar reaction using $\Delta(-)-[Co(CO_3)(en)_2]ClO_4$ gave equal amounts of the $\Delta(-)$ and $\Lambda(+)$ isomers. From these results they ascribed the stereoselective isolation of the $\Lambda(+)$-perchlorate as reported by previous authors to the solubility difference between the isomers. In the experiments according to Legg and Steele, Buckingham et al. obtained equal quantities (50 $\pm$ 3%) of the $\Lambda(+)$ and $\Delta(-)$ isomers. In addition, they observed that mutarotation of the $\Lambda(+)$ isomer in 0.05 mol/dm³ NaOH resulted in 53% $\Lambda(+)-[Co(L-glu)(en)_2]^+$ and 47% $\Lambda(+)-[Co(D-glu)(en)_2]^+$ at the equilibrium.

$[Co(L-Hasp)_2(diamine)]^+$ and $[Co(L-Hasp)(diamine)_2]^{2+}$: The stereochemistry of the mixed L-aspartato-diamine complexes was studied by Kojima and Shibata, using R-propylenediamine [28], ethylenediamine [29], and trimethylenediamine [30]; the complexes were prepared from the reaction between $trans-[CoCl_2(diamine)_2]^+$, L-H$_2$asp, and activated charcoal at pH ca. 10 and 55 °C. Ion-exchange chromatography gave the following results: Where the diamine is R-pn or en, the amount formed in the reaction system decreases in the order of

$$cis(N)trans(O)-\Lambda > cis(N)cis(O)-\Delta > cis(N)trans(O)-\Delta \gtrsim$$
$$trans(N)cis(O)-\Lambda > cis(N)cis(O)-\Lambda \gg trans(N)cis(O)-\Delta .$$

Where the diamine is tn, the order is

$$cis(N)cis(O)-\Delta > cis(N)trans(O)-\Lambda \simeq trans(N)cis(O)-\Lambda >$$
$$cis(N)cis(O)-\Lambda > cis(N)trans(O)-\Delta > trans(N)cis(O)-\Delta .$$

Throughout the three different diamine complexes, the Λ in $cis(N)trans(O)$, the Δ in $cis(N)cis(O)$, and the Λ isomers in $trans(N)cis(O)$ are preferably formed.

For $[Co(L-Hasp)(R-pn)_2]^{2+}$ eight stereoisomers are possible when the orientation of the coordinating R-pn and the Δ and Λ configurations are considered. Experimentally four isomers of the Λ and two isomers of the Δ configuration were isolated by ion-exchange chromatography with the distribution total-Λ/total-Δ = 49/51. The isomer distribution in $[Co(gly)(R-pn)_2]^{2+}$ had been found to be Λ/Δ = 13/87; the marked stereoselectivity due to equatorial orientation of the CH$_3$ groups on the R-pn rings was observed in the Δ isomers [31]. This is why the workers explained the lack of stereoselectivity by the cancellation of the preferred formation of Δ form due to the chelating R-pn with the preferred formation of Λ form due to the chelating L-aspartate. The isomer distribution in $[Co(L-Hasp)(en)_2]^{2+}$ was found to be Λ/Δ = 60/40. The same result had been obtained with a different reaction system [26]. Evidence for an NH$_2$ group forming a hydrogen bond with the β-carboxylate of chelated aspartate was found in the PMR spectrum of the NH$_2$ protons for $\Lambda-[Co(L-asp)(en)_2]^+$ in D$_2$O.

[Co(sar)(en)₂]²⁺ : For the bis(ethylenediamine)sarcosinatocobalt(III) complex, there are four possible isomers due to the asymmetric configurations about both the Co(III) ion and the sarcosinato N atom. Figure 4.5 shows two of the isomers, Δ-S and Δ-R; the latter appears to be less stable than the former due to nonbonded repulsive interaction between the hydrogen atoms on the CH₃ group and those on the adjacent en ring.

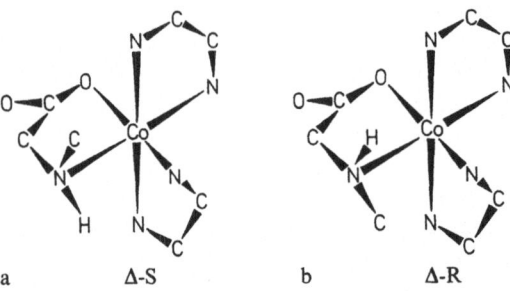

a   Δ-S   b   Δ-R

Fig. 4.5a. and b. Structures of Δ-[Co(S-sar)(en)₂]²⁺ **a** and Δ-[Co(R-sar)-(en)₂]⁺ **b**. Some of H atoms attached to C and N atoms are omitted (from Ref. [36])

In 1966 Backingham et al. [32] repeated the classical resolution [33] of this complex and established that the configuration about the N center of the chelated sarcosinate is controlled by the configuration about the metal ion. Only the Δ-S and Λ-R isomers were formed both under preparative and equilibrium conditions. Based on stereomodels and considering nonbonded interaction, they calculated from the strain energy that the Δ-R species was less stable by *ca.* 8 kcal/mol than the Δ-S. Later, Anderson et al. [34] dealt with [Co(en)₂(*N*-Me-(S)-ala)]²⁺ (*N*-Me-(S)-ala = *N*-methyl-(S)-alaninate) by considering the Δ-R species to be detectable in a concentration of ∼5% by PMR, ion-exchange, or polarimetric methods. However, careful ion-exchange chromatography of equilibrated solutions of the two complexes using Dowex 50W-X2 and Sephadex SP-25 cation-exchangers and NaClO₄—HCl (pH 4) or Na₂HPO₄—NaH₂PO₄ (pH 6.2) eluent failed to separate the isomers. The actual ΔG difference therefore probably exceeds ∼2 kcal/mol.

Fujita et al. [35,36], however, isolated all four isomers by SP-Sephadex chroamtography eluting with sodium (+)-tartratoantimonated(III). They obtained Δ(—)₅₈₉-S, Λ(+)₅₈₉-R, Δ(—)₅₈₉-R and Λ(+)₅₈₉-S, the latter two being the less stable pair. The Δ-S isomer in aqueous solution showed mutarotation, indicating epimerization about the sarcosinato N atom. Equilibration experiments at 25 °C resulted in 84.8% of Δ-S and 15.2% of Δ-R; from this result the ΔG difference was estimated to be *ca.* 1.0 kcal/mol.

4.5 Dipeptide Complexes

Complexes of cobalt with dipeptides from several laboratories led to confusion about the variety of the species formed and their structure [37–40]. In 1966, Gillard et al. [41] prepared several salts of bis(glycylglycinato)cobaltate(III), [Co(glygly)₂]⁻, by treating an aqueous solution (pH 6.5∼9.0) containing cobalt(II) ion and glycylglycine with oxygen. The same anionic complex was obtained from glycylglycine with hexaamminecobalt(III) chloride or sodium tricarbonatocobaltate(III). The X-ray crystal structure of $NH_4[Co(glygly)_2] \cdot 2\,H_2O$ established that the complex anion contains two planar terdentate glycylglycinate ligands, two coordinated oxygen atoms being in the *cis* position [42]. The partial resolution of this complex anion was carried out on a column of potato starch by elution with 30% ethanol—water [43]. The more strongly adsorbed (+)-enantiomer showed a dominant positive CD peak under the first absorption band, the enantiomer being assigned to the $\Lambda(+)$-configuration of Fig. 4.6.

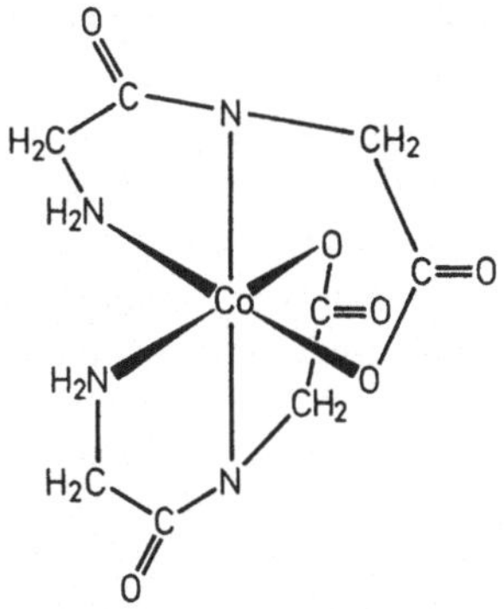

Fig. 4.6. The configuration of $\Lambda(+)$-$[Co(glygly)_2]^-$ (from Ref. [43])

When a dipeptide has one or two L-amino acid residues, the bisdipeptide complex of a metal gives rise to diastereoisomers in unequal amounts. McKenzie [44] investigated the stereoselectivity in the bis-complexes of glycyl-L-leucine, L-alanyl-glycine and L-alanyl-L-alanine. The purple-red aqueous solutions of the anions $[Co(gly\text{-}L\text{-}leu)_2]^-$, $[Co(L\text{-}ala\text{-}gly)_2]^-$ and $[Co(L\text{-}ala\text{-}L\text{-}ala)_2]^-$, prepared from the peptide with $Na_3[Co(CO_3)_3]$ $3\,H_2O$ in water/activated charcoal; all showed very strong negative Cotton effects in their ORD spectra, suggesting that the reaction solutions contained mainly only one of the two diastereoisomers. In order to determine the formation ratio, chromatographic separation was tried for the reaction mixture of $[Co(gly\text{-}L\text{-}leu)_2]^-$ on an alumina column with aqueous methanol. All fractions gave negative Cotton effects of the same shape and intensity as those of the original reaction mixture, indicating that none, or very little, of the second diastereoisomer was separable. The possible source of such marked stereoselectivity was stated as follows: "The molecular models show that the possibilities of hydrogen-bonding between the compound and solvent water molecules are very different for the two diastereoisomers and this is the origin of the observed stereoselectivity."

Stadtherr and Martin [45] investigated stereoselectivity by proton magnetic resonance spectra of solutions prepared by aerating a 2:1 molar ratio mixture of peptide and cobalt(II) salt; the relative percentages of the produced diastereoisomers were evaluated from the peak hights due to methyl or methylene-methin portion. For the product of $[Co(gly\text{-}L\text{-}ala)_2]^-$ formed over a period of days at pH 9.5, the ratio of the isomers was found to be 58/42, for $[Co(L\text{-}val\text{-}gly)_2]^-$ and $[Co(L\text{-}pheala\text{-}gly)_2]^-$ *ca.* 66/33 and *ca.* 70/30, respectively. The stereoselectivity may originate in a binuclear peroxo-cobalt(III) intermediate.

Boas et al. [46] prepared and separated the meridional isomers of the $[Co(\alpha_1-\alpha_2)_2]^-$ type complexes ($\alpha_1-\alpha_2$ represents the dianion of the dipeptide $H_2\alpha_1-\alpha_2$, where α_1 is the N-terminal residue). Seven methods were used to prepare bis(dipeptidato)-cobaltate(III) complexes; (i) by oxygenation of cobalt(II), (ii) from cobalt(II) carbonate, (iii) from sodium tricarbonatocobaltate(III), (iv) from hexaamminecobalt-(III) chloride, (v) from hexakis(urea)cobalt(III) perchlorate, (vi) from cobalt(III) hydroxide oxide, (vii) from triammine(glycylglycinato)cobalt(III). The methods starting from Co^{II} gave more minor products, $Na_3[Co(CO_3)_3] \cdot 3\,H_2O$ gave less, and cobalt(III) hydroxide oxide gave very little of these products. Thus, the method (vi) was prefered. The peptides used were gly-gly, L-ala-gly, gly-L-ala, L-leu-gly, gly-L-leu, L-phe-gly, gly-L-phe, L-ala-L-ala, L-ala-D-ala, L-leu-L-leu.

Two diastereoisomers of each bis(dipeptidato) complexes were separated by anion-exchange chromatography on QAE-Sephadex A-25, and their absolute configurations were determined by PMR, electronic, circular-dichroism spectroscopy. Referring to the stereoselectivity, the workers said that with optically active peptides the formation is stereoselective, but not stereospecific as had earlier been suspected.

4.6 Tris-Complexes of 3-Substituted Camphor

A question arises whether stereoselectivity would be shown in tris complexes with ligands forming planar chelate rings. One group of such ligands is the 3-substituted camphors (**I**).

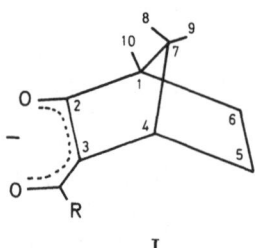

I

The formyl derivative (R = H) is hydroxymethylenecamphor (Hhmc) and the acetyl derivative (R = CH_3) is 3-acetyl camphor (Hatc). For a tris(3-substituted camphorato)cobalt(III) complex, four diastereoisomers are possible; Δ-*mer*, Λ-*mer*, Δ-*fac* and Λ-*fac* (Fig. 4.7).

In 1966, Dunlop et al. [47] reported on the tris-complexes of (+)-hydroxymethylene-camphorate, $[Co\{(+)-hmc\}_3]$; they prepared this complex by the reaction of $Na[Co(CO_3)_3] \cdot 3\,H_2O$ and the ligand in a water—benzene mixture (see 2 · 4) and separated the product into the Λ-*mer* and Λ-*fac* isomers in the ratio 85/15. This marked stereoselectivity was confirmed by similar studies on the $[Co\{(-)-hmc\}_3]$ and $[Co\{(+)-atc\}_3]$ complexes [48]. For the former complex, Δ-*mer* and Δ-*fac* isomers

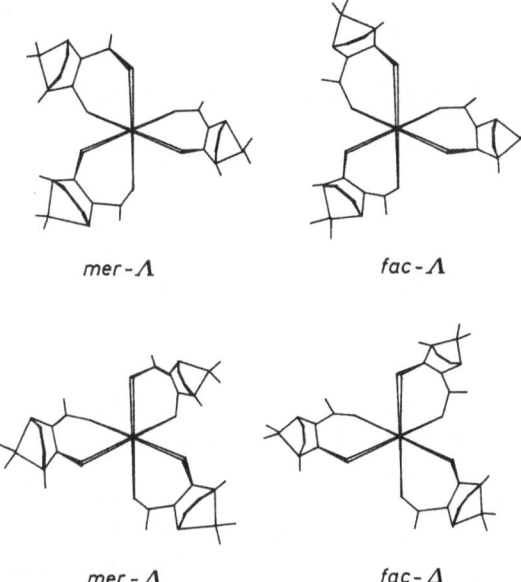

mer - Λ   fac - Λ

mer - Δ   fac - Δ

Fig. 4.7. The four possible stereo-isomers of the tris(3-substituted camphorato)cobalt(III) complex (from Ref. [48])

were separated by chromatography on alumina, greater abundance being found in the Δ-*mer* isomer. The stereoselective effects of (+)-atc ligand appeared to be similar to those of (+)-hmc.

However, Springer et al. [49] found four isomers of [Co{(+)-atc}₃] showing only slight stereoselectivity; the complex was carefully chromatographed on a column of acid-washed alumina and the four isomers were characterized by PMR, ORD and CD spectra. Separately, in equilibration experiments a solution of Δ-*mer* isomer in benzene was sealed in PMR tube in *vacuo*, and the solution was heated to 60 °C until the PMR spectrum, monitored on the thermally quenched sample in intervals, changed no more. The spectrum of the resulting mixture was analyzed for the relative amounts of the isomers. The mole percentages of the isomers were 45.2, 31.3, 16.4 and 7.0 % for Λ-*mer*, Δ-*mer*, Λ-*fac* and Δ-*fac*, respectively. In a preparative reaction mixture the stereoselectivity for Λ-*mer*/Δ-*mer* = 1.4 and Λ-*fac*/Δ-*fac* = 2.3 indicated a slight stereoselectivity in favour of the Λ isomers.

At about the same time, King and Everett [50] also separated the four isomers of [Co{(+)-atc}₃] by preparative thin-layer chromatography on silica gel. The relative amounts of the diastereomers were determined from each TLC band, dissolved in an organic solvent. The absorbances of the solutions resulted in 48.2, 29.5, 12.9 and 9.4 % for Λ-*mer*, Δ-*mer*, Λ-*fac* and Δ-*fac*, respectively. The stereoselectivities, Λ-*mer*/Δ-*mer* and Λ-*fac*/Δ-*fac* were calculated as 1.63 and 1.37, respectively.

In this connection, complexes of hydroxymethylenecarvone (Hhmcar), **II**, and hydroxymethylenepulegone (Hhmpul), **III** were prepared [51]; the relative abundance in the [Co{(−)-hmcar}₃] isomers were 48, 26, 18 and 8 % for Λ-*mer*, Δ-*mer*, Δ-*fac* and Λ-*fac*, respectively; the stereoselectivity was calculated for Λ-*mer*/Δ-*mer* = 1.85 and Λ-*fac*/Δ-*fac* = 0.44. The observed stereoselectivity was stated to result from different interligand interactions involving substituents on the asymmetric carbons, although such interactions are not readily visible in space-filling models.

II   III

4.7 Complexes Containing Optically Active Tartaric Acid

Haines et al. [52] investigated the stereoselective formation of bis(ethylenediamine)-cobalt(III) complexes containing optically active tartaric acid; from a reaction of $[Co(CO_3)(en)_2]Cl$ and S($-$)-tartaric acid in water at steam-bath temperature, two optically active complexes, Λ-$[Co(S\text{-}C_4H_4O_6)(en)_2]$ Cl and Λ-$[Co(S\text{-}C_4H_3O_6)(en)_2]$, were isolated by TLC with silica gel. In the former complex, the tartrate ion acts as a bidentate chelate agent with one free carboxylic acid (**IV**), and in the latter the tartrate is trinegative.

IV

Similar results were reported on bis(phenanthroline) complexes containing tartaric and malic acids [53]: From a reaction of $[Co(CO_3)(phen)_2]Cl \cdot 5\,H_2O$ and R($+$)-tartaric acid in the dark at room temperature, Λ-$[Co(R\text{-}tart)(phen)_2]ClO_4 \cdot 3\,H_2O$ was isolated by cation-exchange chromatography with Dowex 50W-X8 resin. The corresponding R($+$)-malic complex, Λ-$[Co(R\text{-}malato)(phen)_2]ClO_4 \cdot 2\,H_2O$ was isolated in a similar way.

Tatehata [54] isolated a pair of diastereoisomers of R-tartratobis(phenanthroline) complex; from a reacted solution of cis-$[CoCl_2(phen)_2]Cl \cdot 3\,H_2O$ and sodium R($+$)-tartrate in water at 60 °C, Δ-$[Co(R\text{-}tart)(phen)_2]ClO_4 \cdot 3\,H_2O$ and Λ-$[Co(R\text{-}tart)(phen)_2]ClO_4 \cdot 2\,H_2O$ were isolated with the aid of SP-Sephadex C-25 chromatography. The formation ratio of Δ-R:Λ-R = 43:57 was found. The worker followed the reaction of $[Co(CO_3)(phen)_2]Cl \cdot 5\,H_2O$ and R($+$)-tartaric acid and separated the reaction product into two diastereoisomers, whose formation ratio was found to be Δ-R:Λ-R = 27.6:72.4.

4.8 References

1. Dwyer, F. P., Mellor, D. P. (ed.): Chelating Agents and Metal Chelates. New York and London: Academic Press 1964, p. 199
2. Dunlop, J. H., Gillard, R. D.: Advan. Inorg. Chem. Radiochem. *9*, 185 (1966)
3. Pfeiffer, P., Gassman, T.: Liebig's Ann. Chem. *345*, 45 (1906)
4. Tschugaeff, L., Sokoloff, W.: Ber. *42*, 55 (1909)
5. Smirnoff, A. P.: Helv. Chim. Acta *3*, 177, 194 (1920)
6. Jaeger, F. M., Blumendal, H. B.: Z. Anorg. Chem. *175*, 161, 198, 200, 220 (1928)
7. Corey, E. J., Bailar, J. C., Jr.: J. Am. Chem. Soc. *81*, 2620 (1959)
8. Dwyer, F. P., Garvan, P. L., Shulman, A.: J. Am. Chem. Soc. *81*, 290 (1959)
9. Dwyer, F. P., MacDermott, T. E., Sargeson, A. M.: J. Am. Chem. Soc. *85*, 2913 (1963)
10. Dwyer, F. P., Sargeson, A. M., James, L. B.: J. Am. Chem. Soc. *86*, 590 (1964)
11. Denning, R. G., Piper, T. S.: Inorg. Chem. *5*, 1056 (1966)
12. Dunlop, J. H., Gillard, R. D.: J. Chem. Soc. *1965*, 6531
13. Gillard, R. D., Payne, N. C.: J. Chem. Soc. A *1969*, 1197
14. Gillard, R. D., Laurie, S. H., Price, D. C., Phipps, D. A., Weick, C. F.: J. C. S. Dalton *1974*, 1385
15. Kawasaki, K., Yoshii, J., Shibata, M.: Bull. Chem. Soc. Jpn. *43*, 3819 (1970)
16. Kawasaki, K., Shibata, M.: Bull. Chem. Soc. Jpn. *45*, 3100 (1972)
17. Matsuda, T., Okumoto, T., Shibata, M.: Bull. Chem. Soc. Jpn. *45*, 802 (1972)
18. Matsuda, T., Shibata, M.: Bull. Chem. Soc. Jpn. *46*, 3104 (1973)
19. Takenaka, H., Shibata, M.: Bull. Chem. Soc. Jpn. *49*, 2133 (1976)
20. Ogata, Y., Fujinami, S., Shibata, M.: Sci. Rep. Kanazawa Univ. *25*, 73 (1980)
21. Takeuchi, M., Shibata, M.: Bull. Chem. Soc. Jpn. *47*, 2797 (1974)
22. Dunlop, J. H., Gillard, R. D., Payne, N. C., Robertson, G. B.: Chem. Comm. *1966*, 874
23. Dunlop, J. H., Gillard, R. D., Payne, N. C.: J. Chem. Soc. A *1967*, 1469
24. Gillard, R. D., Payne, N. C., Robertson, G. B.: J. Chem. Soc. A *1970*, 2579
25. Gillard, R. D., Maskill, R., Pasini, A.: J. Chem. Soc. A *1971*, 2268
26. Legg, I. L., Steele, J.: Inorg. Chem. *10*, 2177 (1971)
27. Buckingham, D. A., Dekkers, J., Sargeson, A. M., Marzilli, L. G.: Inorg. Chem. *12*, 1207 (1973)
28. Kojima, Y., Shibata, M.: Inorg. Chem. *10*, 2382 (1971)
29. Kojima, Y., Shibata, M.: Inorg. Chem. *12*, 1000 (1973)
30. Kojima, Y.: Bull. Chem. Soc. Jpn. *48*, 2033 (1975)
31. Kojima, Y., Shibata, M.: Inorg. Chem. *9*, 238 (1970)
32. Buckingham, D. A., Mason, S. F., Sargeson, A. M., Turnbull, K. R.: Inorg. Chem. *5*, 1649 (1966)
33. Meisenheimer, J., Angermann, L., Holsten, H.: Ann. *438*, 261 (1924)
34. Anderson, B. F., Buckingham, D. A., Gainsfold, G. J., Robertson, G. B., Sargeson, A. M.: Inorg. Chem. *14*, 1658 (1975)
35. Fujita, M., Yoshikawa, Y., Yamatera, H.: Chem. Lett. *1976*, 959
36. Fujita, M., Yoshikawa, Y., Yamatera, H.: Bull. Chem. Soc. Jpn. *50*, 3209 (1977)
37. Gilbert, J. B., Otey, M. C., Price, V. E.: J. Biol. Chem. *190*, 377 (1951)
38. Cagliotti, V., Silvestroni, P., Furlani, C.: J. Inorg. Nucl. Chem. *13*, 95 (1960)
39. Beck, M. T., Gorog, S.: Acta Chim. Acad. Sci. Hung. *29*, 401 (1961)
40. Tang, P., Li, N. C.: J. Am. Chem. Soc. *86*, 1293 (1964)
41. Gillard, R. D., McKenzie, E. D., Mason, R., Robertson, G. B.: Coord. Chem. Rev. *1*, 263 (1966)
42. Gillard, R. D., McKenzie, E. D., Mason, R., Robertson, G. B.: Nature *209*, 1347 (1966)
43. Gillard, R. D., Harrison, P. M., McKenzie, E. D.: J. Chem. Soc. A *1967*, 618
44. McKenzie, E. D.: J. Chem. Soc. A *1969*, 1656
45. Stadtherr, L. G., Martin, R. B.: Inorg. Chem. *12*, 1810 (1973)
46. Boas, L. V., Evans, C. A., Gillard, R. D., Mitchell, P. R., Phipps, D. A.: J. C. S. Dalton *1979*, 582
47. Dunlop, J. H., Gillard, R. D., Ugo, R.: J. Chem. Soc. A *1966*, 1540
48. Chen, Y. T., Everett, G. W., Jr.: J. Am. Chem. Soc. *90*, 6660 (1968)

49. Springer, C. S., Jr., Siever, R. E., Geibush, B.: Inorg. Chem. *10*, 1242 (1971)
50. King, R. M., Everett, G. W., Jr.: Inorg. Chem. *10*, 1237 (1971)
51. Everett, G. W., Jr., Chen, L. T.: J. Am. Chem. Soc. *92*, 508 (1970)
52. Haines, R. A., Kipp, E. B., Reimer, M.: Inorg. Chem. *13*, 2473 (1974)
53. Haines, R. A., Bailey, D. W.: Inorg. Chem. *14*, 1310 (1975)
54. Tatehata, A.: Inorg. Chem. *15*, 2086 (1976)

5 Design of Low Symmetry Complexes

5.1 Introduction

From the preceeding chapters it can be seen that the preparative studies of today vely on various techniques of chromatography and optical resolution. It is also important to make a good choice of the conditions to be used; in this sense the reactions should be designed for the object.

Recent applications of the ligand field theory to the transition metal complexes give logical interpretations of d-d absorption spectra, and the theoretical treatments help to predict band positions in the absorption spectra of complexes. On the other hand, today low-symmetry complexes are synthesized which contain ligands of more than three kinds. Their spectral data are expected to further develop theoretical fields of spectroscopy.

This chapter deals mainly with the complexes which are preparable from "the green solution" of tricarbonatocobaltate(III) by diamminedicarbonato-type complexes, and whose absorption or CD spectra are characteristic for low symmetry.

5.2 Complexes Exhibiting Marked Splitting in the Second Absorption Bands

5.2.1 Complexes with CoN_3O_3 Chromophores

The most familiar complexes with CoN_3O_3 chromophores are tris(α-amino-acidato) complexes having two geometrical isomers, α or *mer* and β or *fac*. Basolo et al. [1] compared the solution spectra of the α- and β-[Co(gly)$_3$] with those of known *trans* and *cis* isomers of complexes such as $[CoF_2(en)_2]^+$ and $[CoCl(NO_2)(en)_2]^+$, and assigned the α and β isomers to *trans,cis* and *cis,cis* configurations, respectively, since the *cis,cis* is more alike to an octahedral complex and should have a more symmetrical absorption band, while the *trans,cis* isomer should have a broad band at a lower energy with a lower molar extinction coefficient. Shimura and Tsuchida [2] came to the same conclusion by comparing the solution spectra of the α- and β-[Co(gly)$_3$] with those of *trans*- and *cis*-$[Co(C_2H_3O_2)_2(NH_3)_4]^+$, since the α isomer exhibiting an indication of splitting of the first absorption band is assignable to the *trans,cis* structure. The same applied to tris(alaninato)cobalt(III) [3].

In 1958, Yamatera [4] studied shifting and splitting of the first and the second absorption band due to substitution of ligands. Thus, he supported that a broad and unsymmetrical absorption band observed in the spectrum of an α-isomer is attributable to the splitting of $^1T_{1g}$ state due to the rhombic ligand field in the *trans,cis* configuration, while the sharp symmetrical absorption band of a β-isomer is due to the approximately cubic ligand field in the *cis,cis* configuration.

Carbonatoethylenediamineglycinatocobalt(III), [Co(CO₃)(gly)(en)], also having a CoN₃O₃ chromophore, was prepared [5] by air oxidation from the components, then converted to [CoCl(gly)(en)(H₂O)] by warming with conc. HCl, and finally treated with aqueous KHCO₃ solution to obtain a violet and a red isomer. The solution spectra of the two isomers were characteristic for *mer-* and *fac*-[CoN₃O₃] complexes.

The closely related diamminecarbonatoglycinatocobalt(III), [Co(CO₃)(gly)(NH₃)₂], has three possible geometrical isomers (see Fig. 3.10). They were isolated [6] by column chromatography (see 3.3) and characterized by absorption, PMR and IR spectra. The absorption spectrum of the *mer(trans)* isomer (Fig. 5.1) has a shoulder in the second absorption band region, and no splitting is recognized in the first absorption band. The shoulder disappeared when the solution was acidified with aqueous perchloric acid to give the diaqua complex; when the acidified solution was realkalized with potassium hydrogencarbonate, the spectrum returned to that of the original isomer, indicating configuration retention between the carbonato complex and diaqua complex species.

Further studies revealed: When α-alanine or valine was used in stead of glycine, the isolated *mer(trans)* isomer of [Co(CO₃)(ala or val)(NH₃)₂] had a clear shoulder in the second absorption band, while when β-alanine was used, the isolated *mer-(trans)* isomer of [Co(CO₃)(β-ala)(NH₃)₂] had a similar spectrum to that of the *mer(cis)* isomer [7]. Two *mer* isomers of the diammineglycinatooxalato complex, [Co(ox)(gly)(NH₃)₂], derived from an isomer of [Co(CO₃)(gly)(NH₃)₂] showed the usual spectra [7]. With *N,N*-bis(2-aminoethylglycinate) or iso-diethylenetriamine-monoacetate (*i*-dtma) as a CoN₃O₃ chromophore, a very similar spectrum was found for *mer*-[Co(CO₃)(*i*-dtma)] to that of *mer(trans)*-[Co(CO₃)(gly)(NH₃)₂] [8]. The corresponding *fac* isomer had been prepared [9] from the reaction between [CoCl₂(*i*-dtma)] and Li₂CO₃, but the "the green solution" of [Co(CO₃)₃]³⁻ yielded

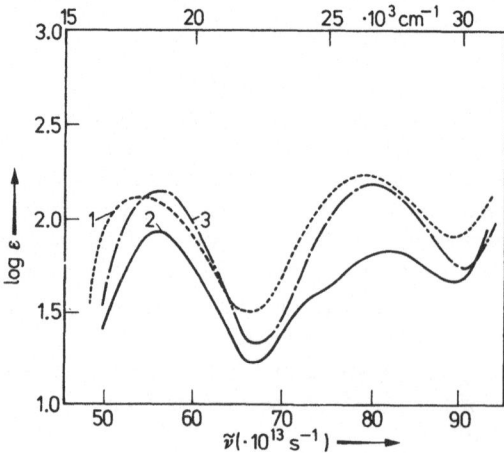

Fig. 5.1. Absorption spectra of the isomers in KHCO₃ aqueous solutions. 1) *mer(cis)*-, 2) *mer(trans)*-, and 3) *fac*-isomer (from Ref. [6])

two isomers. The reaction of the cis-[Co(CO$_3$)$_2$(NH$_3$)$_2$]$^-$ and 2-pyridinecarboxylic acid (picolinic acid, Hpic) resulted in three geometrical isomers of [Co(CO$_3$)(pic)-(NH$_3$)$_2$]. Among them, a violet isomer exhibited a shoulder in the second absorption band region [10].

The absorption spectral data for the complexes with the CoN$_3$O$_3$ chromophore are given in Tabele 5.1. Every complex exhibiting a shoulder in the second absorption band was suggested to have such a configuration that a five-membered N,O-chelate ring and a four-membered carbonate chelate ring lie in the same plane and that the apical positions are occupied by the donor atoms exerting stronger ligand fields than the other donor atoms on the planar positions.

Prior to these works other reports dealt with complexes exhibiting a similar shoulder in the second absorption band; Jørgensen [11] measured the spectra of [Co(OH)(edta)]$^{2-}$ and [Cr(OH)(edta)]$^{2-}$. For the latter complex, Furlani [12] stated such effect to be due to a low symmetry caused by distortion of the octahedron. A similar spectrum was observed [13] for [Co(OH)(penten)]$^{2+}$.

Table 5.1. Absorption Spectral Data for Complexes with CoN$_3$O$_3$ Chromophore

$(\tilde{\nu}/10^3 \text{ cm}^{-1})$

Complexes	$\tilde{\nu}_I(\log \varepsilon)$	$\tilde{\nu}_{II}(\log \varepsilon)$
mer-[co(gly)$_3$]	18.40 (1.99)	26.80 (2.16)
fac-	19.23 (2.20)	26.73 (2.14)
mer-[Co(ala)$_3$]	18.53 (2.00)	26.96 (2.18)
fac-	19.30 (2.27)	26.80 (2.21)
mer(cis)-[CoCO$_3$(gly)(NH$_3$)$_2$]	18.06 (2.13)	26.50 (2.25)
mer(trans)-	18.66 (1.93)	24.0 (sh)
		27.30 (1.85)
fac-	18.73 (2.16)	26.7 (2.22)
mer(trans)-		
[CoCO$_3$(α-ala)(NH$_3$)$_2$]	18.83 (2.00)	24.2 (sh)
		27.33 (2.04)
[CoCO$_3$(val)(NH$_3$)$_2$]	18.76 (1.90)	24.0 (sh)
		27.33 (1.80)
[CoCO$_3$(β-ala)(NH$_3$)$_2$]	18.27 (2.11)	26.83 (2.15)
[Co(ox)(gly)(NH$_3$)$_2$]	18.83 (1.78)	27.33 (1.88)
[Co(gly)(NH$_3$)$_2$(H$_2$O)$_2$]$^{2+}$	18.00 (1.79)	27.60 (1.83)
	19.0 (sh)	
[CoCO$_3$(i-dtma)]	19.36 (2.12)	24.83 (1.7)
		27.43 (1.98)
[CoCO$_3$(pic)(NH$_3$)$_2$]	19.20 (2.04)	25.0 (sh)
		28.4 (sh)

5.2.2 Complexes with CoCN$_3$O$_2$ Chromophores

Based on the working hypothesis on the [CoN$_3$O$_3$]-type complexes, Nakashima and Shibata [14] extended their preparative work to the [Co(CN)(CO$_3$)(N$_3$)]-type complexes, where (N)$_3$ represents three NH$_3$, one NH$_3$ plus one en, or one dien. The possible geometrical isomers are illustrated in Fig. 5.2, where mer and fac refer to

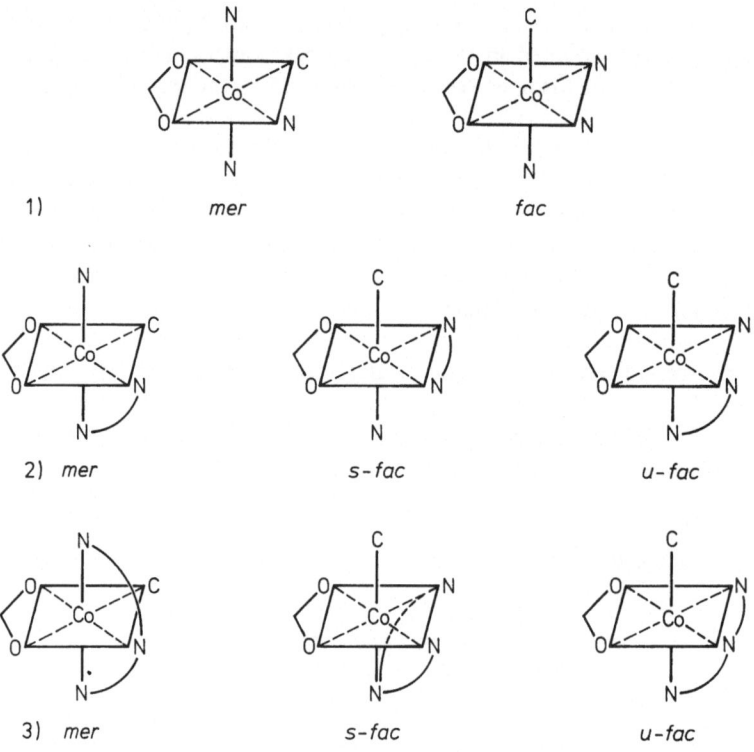

Fig. 5.2. Possible geometrical isomers of the [Co(CN)(O,O)(N)₃]-type complexes; 1) triammine complex, 2) ammineethylenediamine complex, and 3) diethylenetriamine complex

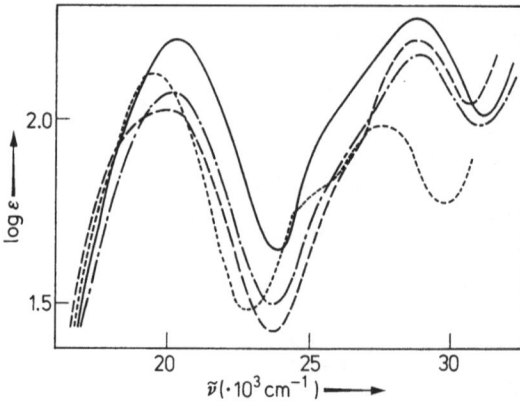

Fig. 5.3. Absorption spectra of *fac*(N)-[Co(CN)(CO₃)(NH₃)₃] (———), *u-fac*(N)-[Co(CN)(CO₃)(NH₃)(en)] (—·—·—), *s-fac*(N)-[Co(CN)(CO₃)(dien)] (————), and *mer*(N)-[Co(CO₃)(*i*-dtma)] (······) (from Ref. [14])

the N atoms, and *s* and *u* are used to distinguish between the isomer having a symmetry plane and that having no symmetry plane. The complexes were prepared by letting potassium cyanide react with "the green solution" at room temperature and then adding ammonia and/or diamine at mild temperatures. The geometrical isomers were separated by column chromatography for neutral complexes.

Both the *mer* and *fac* isomers were obtained for [Co(CN)(CO₃)(NH₃)₃], the *mer* and *u-fac* among the three isomers for [Co(CN)(CO₃)(NH₃)(en)], and the *mer*(N) and *s-fac* among the three isomers for [Co(CN)(CO₃)(dien)].

Each of the *fac* isomers exhibited a shoulder in the second absorption band and no shoulder in the first absorption band (Fig. 5.3). In contrast, the corresponding *fac* isomers of the [Co(CN)(ox)(N)₃] complexes derived from the carbonato complexes exhibited no such shoulder in the second absorption band, but did show a shoulder in the first absorption band (Fig. 5.4).

From the characterized structures, the workers stated that every complex exhibiting a shoulder in the second absorption band has such a configuration that no chelate ring of ethylenediamine backbone is co-planar with the carbonate chelate ring and that the apical positions are occupied by the donor atoms exerting stronger ligand field than the other donor atoms on the planar positions.

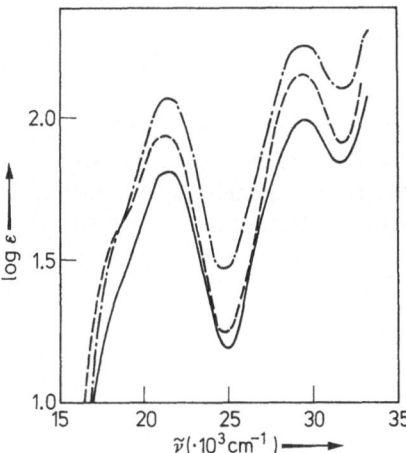

Fig. 5.4. Absorption spectra of *fac*(N)-[Co(CN)-(ox)(NH₃)₃] (– – – –), *u-fac*(N)-[Co(CN)(ox)-(NH₃)(en)] (— · — · —), and *s-fac*(N)-[Co(CN)(ox)-(dien)] (————), in *ca.* 20% H₂SO₄ (from Ref. 14))

5.2.3 Complexes with CoN₂O₄ Chromophores

Most of the complexes with CoN₂O₄ chromophores so far prepared contain the N-donor as amino-acidate, iminodiacetate, or nitrilotriacetate, and the first absorption bands of their *trans*(N) isomers are explicitly split. However, *trans*(N) complexes [15,16] that exhibit splitting not in the first but in the second absorption band were isolated, from the reaction of "the green solution" and pyridine, as crystals of *trans*- and *cis*-[Co(CO₃)₂(py)₂]⁻. From the mixture of the dicarbonato-oxalatocobaltate(III) and pyridine, *trans*-[Co(CO₃)(ox)(py)₂]⁻ was obtained. The reactions of pyridine on aqueous solutions of tris(oxalato)cobaltate(III) and tris-(malonato)cobaltate(III) resulted in *trans*- and *cis*-[Co(ox)₂(py)₂]⁻ and *trans*- and *cis*-[Co(mal)₂(py)₂]⁻, respectively.

The *trans* isomers of the dicarbonato, carbonatooxalato and bis(oxalato) complexes have a split second absorption band (II a and II b); the extent of the splitting is greatest for the dicarbonato complex and smallest for the bis(oxalato) complex (Fig. 5.5). The *trans*(N)-bis(malonato) complex, showed no splitting in either the

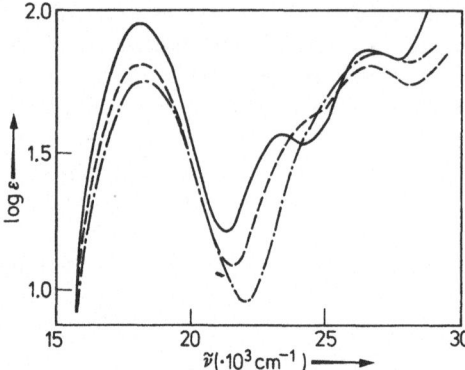

Fig. 5.5. Absorption spectra of *trans* isomers of $[CoCO_3(ox)(py)_2]^-$ (−−−−−), $[Co(ox)_2(py)_2]^-$ (−·−·−), and $[Co(CO_3)_2(py)_2]^-$ (―――) (from Ref. [15])

first or the second absorption band. The absorption spectra are summarized in Table 5.2, which includes the spectral data [17] on some related complexes.

The first band of the three *trans* complexes is expressed by a Gaussian, while the second band is divided into two Gaussians. From a comparison of the spectral data on the first band (*i.e.* T_{1g}) region with those of the related *trans*(N)-$[CoN_2O_4]^-$-type complexes, the workers stated that the observed I band ($18\,200 \sim 18\,400$ cm^{-1}) is mainly due to the $^1A_{1g} \rightarrow {}^1E_g^a$ transition under the tetragonal (D_{4h}) symmetry, the transition to the $^1A_{2g}$ component being perhaps hidden by the foot of the I band.

Table 5.2. Absorption Spectral Data for *trans*-Complexes with CoN_2O_4 Chromophore

trans-Complexes	I Band $\tilde{v}$ (log ε)	II Band $\tilde{v}$ (log ε)
trans-$[Co(CO_3)_2(py)_2]^-$	18.2 (1.95)	23.5 (1.56)
		26.7 (1.86)
trans-$[Co(CO_3)(ox)(py)_2]^-$	18.2 (1.82)	*ca.* 23.8
		26.7 (1.80)
trans-$[Co(ox)_2(py)_2]^-$	18.4 (1.76)	*ca.* 24.2
		26.7 (1.84)
trans-$[Co(mal)_2(py)_2]^-$	18.1 (1.78)	26.0 (1.85)
trans-$[Co(ox)(H_2O)_2(py)_2]^+$	17.9 (1.73)	26.5 (1.95)
trans-$[Co(gly)_2(ox)]^-$	16.70 (1.70)	24.84 (2.23)
	18.87 (2.00)	
trans-$[Co(L\text{-ala})_2(ox)]^-$	16.70 (1.70)	25.97 (2.24)
	18.94 (2.00)	
trans-$[o(L\text{-ser})_2(ox)]^-$	16.70 (1.80)	25.97 (2.36)
	19.05 (2.13)	
trans-$[Co(\beta\text{-ala})_2(ox)]^-$	16.00 (1.76)	26.30 (2.12)
	18.83 (2.00)	
trans-$[Co(ida)_2]^-$	16.67 (1.06)	27.77 (1.75)
	20.40 (1.72)	
trans-$[Co(L\text{-asp})_2]^-$	15.87 (1.16)	26.46 (1.90)
	19.53 (1.90)	

On the basis of a tetragonal (D_{4h}) model, Fujinami et al. [18] developed the procedures for calculating the vibronically induced transition moment and showed a correlation in the band intensity between two nondegenerate transition components, A_2 and B_2. They applied this correlation to the understanding of the absorption spectra of *trans*-$[CrClF(NH_3)_4]^+$ and *trans*-$[CrF_2(NH_3)_4]^+$, which are split in both the first and second absorption bands [19,20]. Furthermore, they explained the spectra of *trans*-$[Co(ox)_2(py)_2]^-$ and *trans*-$[Co(CO_3)_2(py)_2]^-$: The ratio of the oscillator strengths of the $^1A_{2g}$ transition, $f(A_2)$, to that of the $^1B_{2g}$ transition, $f(B_2)$, for cobalt(III) complexes was expressed by

$$\frac{f(A_2)}{f(B_2)} = 1.5 \frac{E(A_2)}{E(B_2)} \times \frac{E(CT) - E(B_2)}{E(CT) - E(A_2)}$$

where $E(A_2)$, $E(B_2)$, and $E(CT)$ denote the energies associated with the respective transitions. Using this equation, the intensity of the $^1A_{1g} \rightarrow {}^1A_{2g}$ transition was evaluated for *trans*-$[Co(ox)_2(py)_2]^-$; the $E(A_{2g})$ value was estimated to be *ca.* 17 000 cm^{-1} in view of the data on *trans*(N)-$[Co(ida)_2]^-$ and *trans*(N)-$[Co(ox)(gly)_2]^-$ (Table 5.2). The $E(CT)$ value was assumed to be 41 000 cm^{-1}, corresponding to the CT maximum for $[Co(ox)_3]^{3-}$. Thus, the ε_{max} value due to the $^1A_{1g} \rightarrow {}^1A_{2g}$ transition was calculated to be *ca.* 10. Similar calculation gave $\varepsilon_{max} = $ *ca.* 16 for *trans*-$[Co(CO_3)_2$-$(py)_2]^-$. These values were much smaller than the observed ε_{max} values in the first absorption band region ($\varepsilon_{max} = 57.5$ for the bis(oxalato) complex and $\varepsilon_{max} = 89.1$ for the dicarbonato complex). Thus, while two split bands were observed in the $^1T_{2g}$ region, only one band assignable to E^a was observed in the $^1T_{1g}$ region. This phenomenon could be rationalized by the relative intensities of the nondegenerate A_2 and B_2 bands.

5.2.4 Complexes Containing Aminoalcoholate Ion

Aminoalcohols such as 2-aminoethanol(ethanolamine, Heta) and 2-amino-1-propanol(propanolamine, Hpra) have been known to combine as a neutral bidentate or as a deprotonated aminoalcoholate [21~23]. In 1973, Ogino et al. [24] communicated CD spectral studies of $[Co(en)_2(Heta)]^{3+}$ and $[Co(en)_2(L-Hpra)]^{3+}$; the complexes were prepared from *trans*-$[CoBr_2(en)_2]^+$ and the ligands *via* $[CoBr(en)_2(Ham)]^+$ (Ham = Heta or L-Hpra) according to Buckingham et al. [25]. Each complex was separated into two diastereoisomers by Sephadex chromatography.

The absorption spectra and the CD spectra of these complexes are distinctively different from those of the corresponding deprotonated complexes produced under alkaline condition. In the absorption spectra of the deprotonated species, $[Co(en)_2$-$(Oam)]^{2+}$ (Oam = deprotonated aminoalcoholate), a shoulder band was observed at *ca.* 25 000 cm^{-1}, and in the CD spectra, an intense band was observed at *ca.* 25 000 cm^{-1} (Fig. 5.6). Nishide et al. [26] extended the work to various complexes of the types $[Co(NH_3)_4(Ham)]^{3+}$, $[Co(en)_2(Ham)]^{3+}$ and $[Co(R-chxn)_2(Ham)]^{3+}$. They observed a shoulder for $[Co(en)_2(Oam)]^{2+}$ at this wavelength, but not for the $[Co(NH_3)_4(Oam)]^{2+}$ complexes. On the other hand, the CD spectra of both complexes exhibited a strong CD band at *ca.* 25 000 cm^{-1}. Thus, they concluded that all the deprotonated complexes have an absorption component at *ca.* 25 000 cm^{-1}, even though no obvious shoulder was observed in the absorption spectra as in the

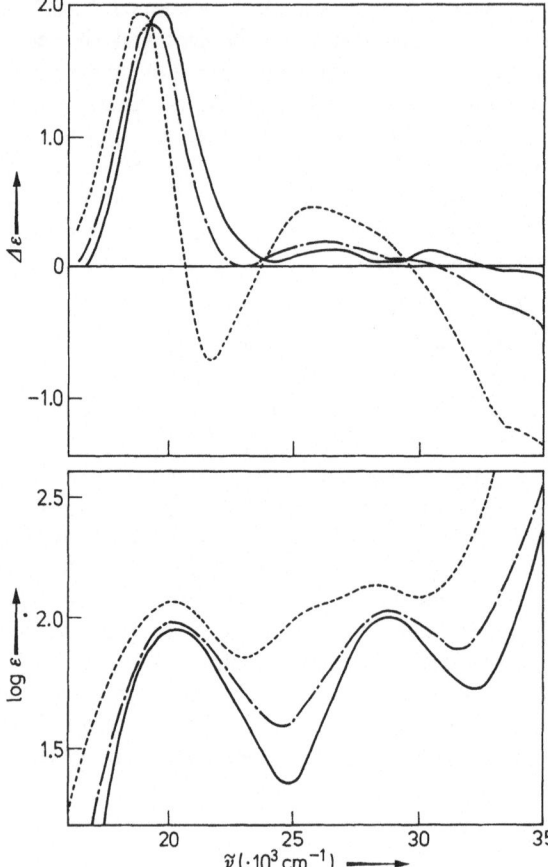

Fig. 5.6. The CD and absorption spectra of $(+)_D$-[Co(en)$_2$(Heta)]-Br$_3 \cdot$ H$_2$O at pH 1.4 (————), pH 2.7 (—·—·—) and pH 9.7 (······) (from Ref. [24])

tetraammine complexes. Furthermore they claimed that the shoulder absorption and characteristic CD band are both related to the coordinated alcoholate oxygen atom with two sets of nonbonding electron pairs.

Okazaki and Shibata [27] reported on the [Co(Heta or S-Hpra)(N)$_2$(O)$_2$]-type complexes in which the (N)$_2$(O)$_2$ moiety denotes two glycinate, two β-alaninate, ethylenediamine plus oxalate, or two ammonia plus oxalate. For a [Co(Heta)-(N)$_2$(O)$_2$]-type complex, the possible geometrical isomers are illustrated in Fig. 5.7,

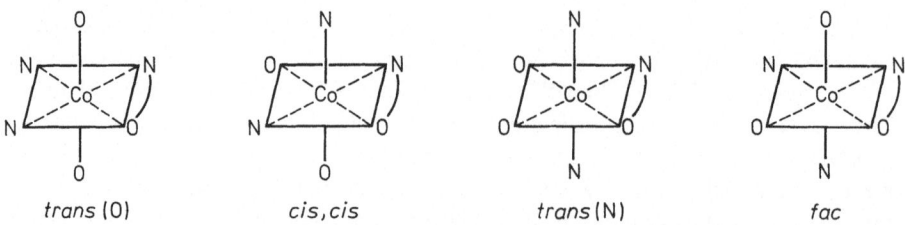

Fig. 5.7. Possible geometrical isomers of [Co(Heta)(N)$_2$(O)$_2$]. N—O denotes Heta

where *trans*(O), *cis,cis* and *trans*(N) are *mer*(N) isomers. When the $(N)_2(O)_2$ moiety is $(gly)_2$ or $(\beta\text{-ala})_2$ four isomers are possible, when $(N)_2(O)_2 = (ox)(en)$ two isomers, *cis,cis* and *fac*, are possible, and when $(N)_2(O)_2 = (NH_3)_2(ox)$ the possible isomers are *cis,cis*, *trans*(N) and *fac*.

These complexes were prepared from carbonato complexes such as $[Co(CO_3)-(am)_2]^-$ (am = gly or β-ala), $[Co(CO_3)(ox)(en)]^-$, *cis*-$[Co(CO_3)(ox)(NH_3)_2]^-$ as the starting materials. For the isolation of the geometrical isomers, each reaction mixture was first chromatographed on a cation-exchange resin in Na^+ form using an aqueous NaCl as eluent, and then rechromatographed under alkaline condition, using water as eluent.

The absorption spectra measured at pH *ca.* 2.0 and pH *ca.* 8.0 are shown in Figs. 5.8 and 5.9, respectively. Each of the deprotonated complexes exhibits a spectrum quite different from that of the corresponding protonated complexes.

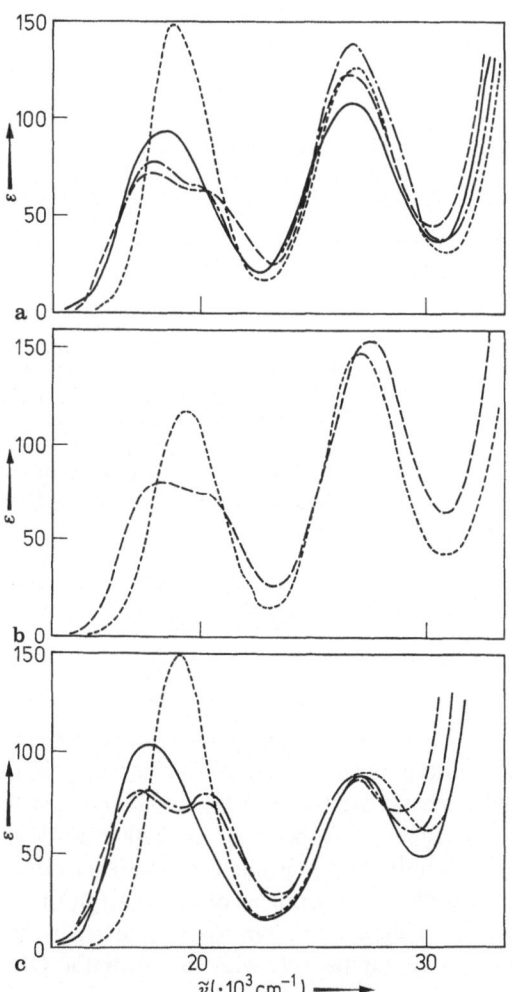

Fig. 5.8a–c. Absorption spectra of protonated complexes, **a** $[Co(gly)_2(Heta)]^+$, **b** $[Co(ox)(en)(Heta)]^+$, and **c** $[Co(\beta\text{-ala})_2(Heta)]^+$; —— *trans*(O), – – – – *cis,cis*, –·–·– *trans*(N), and ······· *fac*. The ε values of *cis,cis*-, *fac*-$[Co(\beta\text{-ala})_2(Heta)]^+$ are taken arbitrarily (from Ref. [27])

$\tilde{\nu}(\cdot 10^3 cm^{-1})$ ⟶

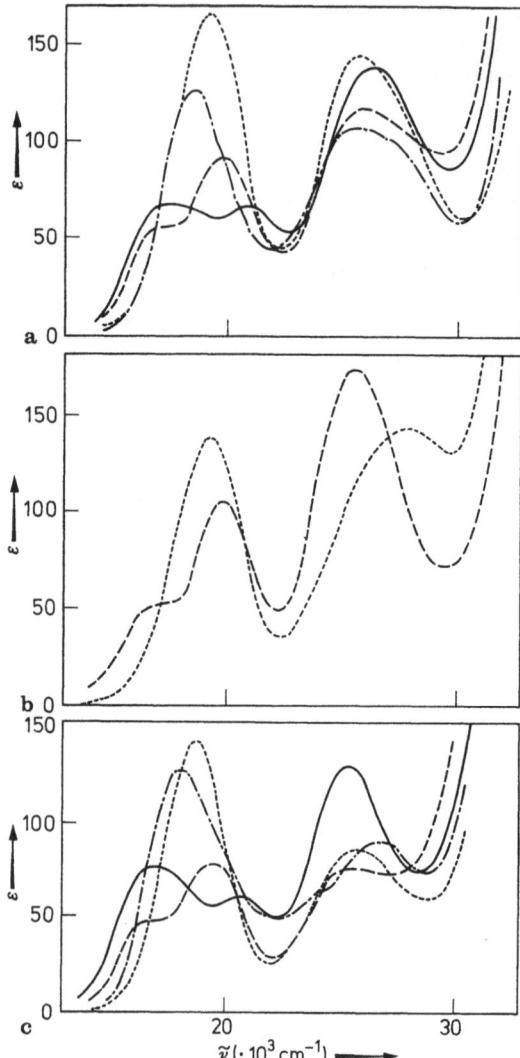

Fig. 5.9a–c. Absorption spectra of deprotonated complexes, **a** [Co(gly)$_2$-(eta)], **b** [Co(ox)(eta)(en)], and **c** [Co(β-ala)$_2$(eta)]; ———**W** *trans*(O), – – – – – *cis,cis*, —·—·— *trans*(N), and ······· *fac*. The ε values of *cis,cis*-, *fac*-[Co(β-ala)$_2$(eta)] are taken arbitrarily (from Ref. [27])

The *trans*(N- isomer of each of [Co(gly)$_2$(eta)] and [Co(β-ala)$_2$(eta)] exhibits recognizable splitting in the T$_{2g}$ band and no splitting in the T$_{1g}$ band. Furthermore, in the spectrum of *cis,cis*-[Co(gly)$_2$(eta)], the T$_{2g}$ band exhibits a broad maximum, and the T$_{1g}$ band shows explicit splitting. The intense CD peak at *ca.* 25000 cm^{-1} was observed in the CD spectrum for *fac*-[Co(gly)$_2$(eta)]. From the polarized crystal spectra and the results of X-ray analysis [28] of the isomers of [Co(eta)(gly)$_2$], the σ- and π-antibonding parameters for the alcoholate O atom were estimated by the Angular Overlap Model [29] and found to be comparable with those for the OH$^-$ ligand.

5.3 Complexes with Optical Activity Due to Unidentate Ligands

5.3.1 *cis,cis*-[Coa$_2$b$_2$CC] and *cis,cis,cis*-[Coa$_2$b$_2$c$_2$]

Most CD spectra have been studied to investigate various sources of optical activity such as distribution of chelate rings, conformation of chelate rings, vicinal effect due to asymmetric carbon in an optically active ligand and vicinal effect due to an asymmetric donor atom [30]. Much less studies have been done on complexes whose optical activity results from the arrangement of unidentate ligands because such chiral complexes could not be synthesized.

Hawkins et al. observed the CD spectrum [31] of an optically active *cis*-diammine-*cis*-dichloro-mono(ethylenediamine)cobalt(III), $(+)_{589}$[CoCl$_2$(NH$_3$)$_2$(en)]$^+$ and prepared it together with related complexes [32]. As the starting material, *cis*-[Co(CO$_3$)-(NH$_3$)$_2$(en)] ClO$_4$ was produced according to Bailar and Peppard [33]. The racemate was then resolved by diastereoisomer formation with ammonium *trans*-diammine-(R-1,2-propanediamine)bis(sulfito)cobaltate(III) [34]. The less soluble diastereoisomeric salt was separated through an anion exchange column (BioRad AG1-X2, ClO$_4^-$ form) in order to obtain $(+)_{589}$[Co(CO$_3$)(NH$_3$)$_2$(en)] ClO$_4$. When dry hydrogen chloride was passed over the finely ground $(+)_{589}$-compound under anhydrous conditions, the aimed dichloro compound was obtained quantitatively.

The absorption and CD spectral data on the isomer derived from the $(+)_{589}$-carbonato complex are cited in Table 5.3 with the data on related complexes. The CD spectrum of the parent $(+)_{589}$[Co(CO$_3$)(NH$_3$)$_2$(en)]$^+$ complex is very similar to that of $(+)_{589}$[Co(CO$_3$)(en)$_2$]$^+$ whose absolute configuration has been known to be Λ. The same configuration was assigned to $(+)_{589}$-[Co(CO$_3$)(NH$_3$)$_2$(en)]$^+$, on the basis of the fact that *cis,cis*-[CoCl$_2$(NH$_3$)$_2$(en)]$^+$ was derived from the $(+)_{589}$-carbonato complex by a solid state reaction without isomerization. Thus, R configuration was assigned to this dichloro complex, where the symbol R is defined according to the method of Cahn et al. [35] (Fig. 5.10). When the CD spectrum of R-[CoCl$_2$(NH$_3$)$_2$(en)]$^+$ was compared with that of $(+)_{589}$[CoCl$_2$(en)$_2$]$^+$, the same sign of Cotton effect was found (Table 5.3). However, Hawkins et al. have mentioned that [30] "there is no *a priori* reason for suggesting that the two should have the same sign of Cotton effect for the comparable transitions, because the two complexes attain their dissymmetry in different ways."

Table 5.3. Absorption and CD Spectral Data of *cis,cis*-[CoCl$_2$(NH$_3$)$_2$(en)]$^+$ and Related Complexes
(T$_{1g}$ Band)

Complex	Absorption		CD	
	$\tilde{\nu}$ (cm^{-1})	ε	$\tilde{\nu}$ (cm^{-1})	$\Delta\varepsilon$
cis,cis-[CoCl$_2$(NH$_3$)$_2$(en)]$^+$ (in DMSO)	18 150	78.7	16 230	−0.408
			18 350	+0.062
$(+)_{589}$[Co(CO$_3$)(NH$_3$)$_2$(en)]$^+$	19 420	119.5	18 750	+1.530
	27 940	121.0	25 500	+0.130
			27 420	−0.052
Λ-$(+)_{589}$[Co(CO$_3$)(en)$_2$]$^+$	19 490	143	18 870	+3.7
$(+)_{589}$[CoCl$_2$(en)$_2$]$^+$	18 690	69	16 260	−0.6
			18 580	+0.7

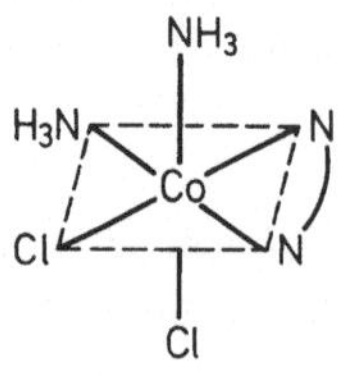

Muraji Shibata

(R)

Fig. 5.10. Absolute configuration of *cis,cis*-[CoCl$_2$(NH$_3$)$_2$(en)]$^+$

Shibata et al. [36)] investigated other complexes, *cis,cis*-diamminecarbonatodicyano-cobaltate(III), *cis,cis*-[Co(CN)$_2$(CO$_3$)(NH$_3$)$_2$]$^-$ and *cis,cis*-diamminediaquaoxalato-cobalt(III), *cis,cis*-[Co(ox)(NH$_3$)$_2$(H$_2$O)$_2$]$^+$ and prepared complexes with *cis,cis*-distributions of unidentate ligands. They [37,38)] also synthesized series of *cis,cis*-[Co(CN)$_2$(O, O)(NH$_3$)$_2$]$^-$ and *cis,cis*-[Co(NO$_2$)$_2$(O, O)(NH$_3$)$_2$]$^-$ complexes (O, O represents CO$_3^{2-}$, ox^{2-}, and mal^{2-}), as well as the closely related *cis*-[Co(CN)$_2$(O, O)-(en)]$^-$ and *cis*-[Co(NO$_2$)$_2$(O, O)(en)]$^-$ complexes. They devised a method depending essentially on the fact that the tricarbonatocobaltate(III) species prefers successive *cis*-substitutions by the desired unidentate ligands. Schemes 1 and 2 represent the pathways of syntheses.

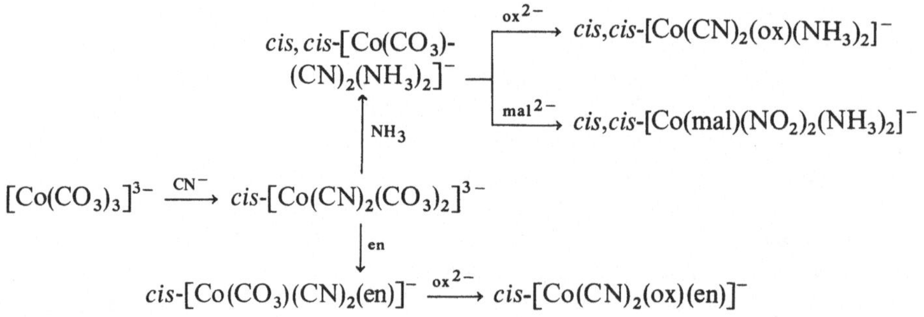

Scheme 1. Syntheses of *cis,cis*- and *cis*-Dicyano Complexes

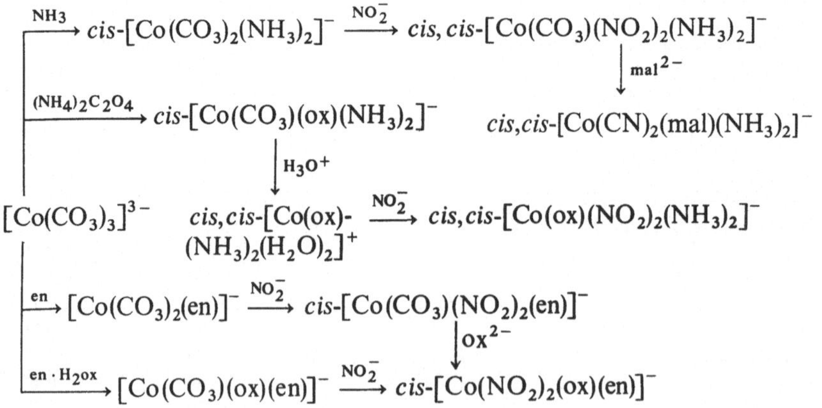

Scheme 2. Syntheses of *cis,cis*- and *cis*-Dinitro Complexes

98

Syntheses of *cis,cis*-[Co(CN)$_2$(CO$_3$)(NH$_3$)$_2$]$^-$ and *cis,cis*-[Co(NO$_2$)$_2$(CO$_3$)(NH$_3$)$_2$]$^+$ are described as examples. The action of KCN on "the green solution" of [Co(CO$_3$)$_3$]$^{3-}$ at room temperature gives a deep red solution containing *cis*-[Co(CN)$_2$(CO$_3$)$_2$]$^{3-}$ as the main product. After neutralization with aqueous HClO$_4$, ammonium perchlorate and concentrated aqueous NH$_3$ are mixed to the solution. After some time at a mild temperature, the concentrated solution is subjected to ion-exchange chromatography (Dowex 1-X8 in Cl$^-$ form, 0.3 mol/dm^3 NaCl eluent) to collect the desired complex species in a fraction of yellow-orange, from which *cis,cis*-Na[Co(CN)$_2$(CO$_3$)(NH$_3$)$_2$] · 2 H$_2$O is obtained.

Another complex, *cis,cis*-[Co(NO$_2$)$_2$(CO$_3$)(NH$_3$)$_2$]$^-$, was prepared from a mixture of *cis*-K[Co(CO$_3$)$_2$(NH$_3$)$_2$] · H$_2$O and KNO$_2$ in water with dil. acetic acid at room temperature. After ion-exchange chromatography (Dowex 1-X8 in Cl$^-$ form, 0.1 mol/dm^3 KCl eluent) the earlier eluate is *trans*(NH$_3$)*cis*(NO$_2$)-isomer and the later is the desired isomer, which is crystallized as K[Co(NO$_2$)$_2$(CO$_3$)(NH$_3$)$_2$] · 0.5 H$_2$O.

The optical resolution of each complex synthesized was achieved by using either (−)$_{589}$[Co(NO$_2$)$_2$(en)$_2$] (C$_2$H$_3$O$_2$) or (−)$_{589}$[Co(ox)(en)$_2$] (C$_2$H$_3$O$_2$) in an amount equivalent to half the amount of the racemate of the complex.

The absorption and CD spectra of the dicyano complexes are cited in Fig. 5.11, which also shows the spectra of *cis,cis,cis*-[Co(CN)$_2$(NH$_3$)$_2$(H$_2$O)$_2$]$^+$ derived from the resolved [Co(CN)$_2$(CO$_3$)(NH$_3$)$_2$]$^-$ complex by acid-hydrolysis. The data of the T$_{1g}$

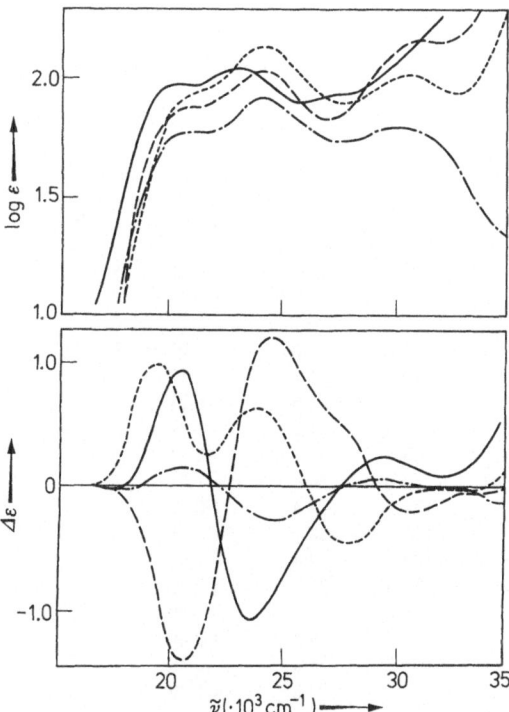

Fig. 5.11. Absorption and CD spectra of *cis,cis*-[Co(CN)$_2$CO$_3$(NH$_3$)$_2$]$^-$ (————), *cis,cis,cis*-[Co(CN)$_2$(H$_2$O)$_2$(NH$_3$)$_2$]$^+$ (—·—·—·—), *cis,cis*-[Co(CN)$_2$ox(NH$_3$)$_2$]$^-$ (— — — —), and *cis,cis*-[Co(CN)$_2$mal(NH$_3$)$_2$]$^-$ (– – – – –) (from Ref. [38])

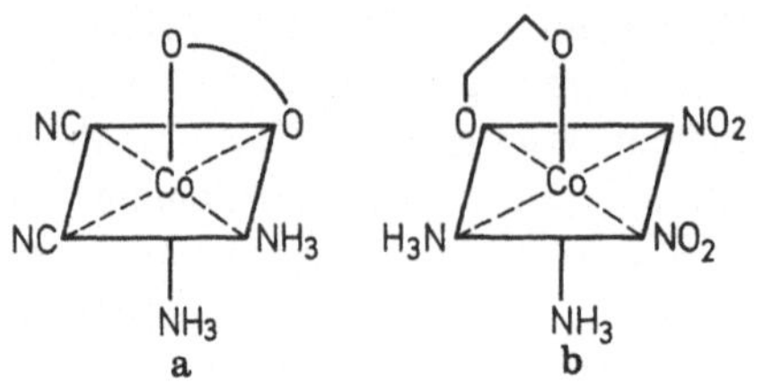

Fig. 5.12a. and b. Absolute configurations of a $S(+)_{589}[Co(CN)_2mal(NH_3)_2]^-$ and b $R(-)_{589}[Co(NO_2)_2ox(NH_3)_2]^-$

band maxima with $\Delta\varepsilon$ are summarized in Table 5.4 for the dicyano complexes and in Table 5.5 for the dinitro complexes.

The absolute configuration was determined by X-ray analysis with $K(+)_{589}[Co(CN)_2(mal)(NH_3)_2] \cdot H_2O.$[39] and $(-)_{589}[Co(NO_2)_2(en)_2](-)_{589}[Co(NO_2)_2(ox)-(NH_3)_2]$ [40] as illustrated in Fig. 5.12. With the other complexes, the workers determined the absolute configurations tentatively by comparing the observed CD spectra in the T_{1g} region with those of the two standards; the results are summarized in Tables 5.4 and 5.5, where symbols R and S are used not only for *cis,cis* complexes, but also *cis*-complexes for the sake of convenience.

While $S(+)_{589}[Co(CN)_2(mal)(NH_3)_2]^-$ shows a $(+, +)$ pattern, $(-)_{589}[Co(CN)_2-(CO_3)(en)]^-$ shows a $(-, -)$ pattern; hence R configuration is assigned to the latter complex. The configuration of $(-)_{589}[Co(CN)_2(en)(H_2O)_2]^+$ is regarded to be the same as that of the parent carbonato complex, because acid-hydrolysis proceeds with retention of configuration. The $(-, +, -)$ pattern of $(+)_{589}[Co(CN)_2(CO_3)-(NH_3)_2]^-$ resembles that of $R(-)_{589}[Co(CN)_2(en)(H_2O)_2]^+$, if the first very weak

Table 5.4. Absorption and CD Spectral Data of Dicyano Complexes and Tentative Absolute Configurations

$(\tilde{v}/10^3 \text{ cm}^{-1})$

Complex	Absorption		CD		absolute configuration
	$\tilde{v}$	ε	$\tilde{v}$	$\Delta\varepsilon$	
$(+)_{589}[Co(CN)_2(CO_3)(NH_3)_2]^-$	20.7	92.0	18.2	−0.02	R
	23.5	104	20.9	+0.98	
			24.1	−1.11	
$(+)_{589}[Co(CN)_2(NH_3)_2(H_2O)_2]^+$	21.3	58.1	18.6	−0.01	R
	24.5	79.4	21.1	+0.17	
			24.6	−0.27	
$(-)_{589}[Co(CN)_2(ox)(NH_3)_2]^-$	21.3	72.5	21.2	−1.44	S
	24.3	103	24.9	+1.24	
$(+)_{589}[Co(CN)_2(mal)(NH_3)_2]^-$	21.5 (*sh*)		20.0	+1.02	S^x
	24.4	132	24.4	+0.68	
$(-)_{589}[Co(CN)_2(CO_3)(en)]^-$	20.9	114	20.7	−0.86	R
	23.1	102	24.4	−1.46	
$(-)_{589}[Co(CN)_2(en)(H_2O)_2]^+$	21.6	75.9	21.1	+0.56	R
	24.3	83.1	25.3	−0.80	
$(+)_{589}[Co(CN)_2(ox)(en)]^-$	21.5	95.5	19.3	+0.08	S
	24.3	104	21.3	−0.18	
			24.8	+1.49	

Table 5.5. Absorption and CD Spectral Data of Dinitro Complexes and Tentative Absolute Configuration

$$(\tilde{v}/10^3 \text{ cm}^{-1})$$

Complex	Absorption		CD		absolute configuration
	$\tilde{v}$	ε	$\tilde{v}$	$\Delta\varepsilon$	
$(-)_{589}[Co(NO_2)_2(CO_3)(NH_3)_2]^-$	20.6	208	18.8	+0.12	S
			21.7	−0.91	
$(+)_{589}[Co(NO_2)_2(NH_3)_2(H_2O)_2]^+$	21.2	161	19.2	−0.12	S
			22.2	+0.23	
$(-)_{589}[Co(NO_2)_2(ox)(NH_3)_2]^-$	20.9	161	19.5	−0.89	R^x
			22.3	+1.47	
$(+)_{589}[Co(NO_2)_2(mal)(NH_3)_2]^-$	20.9	168	18.3	−0.03	S
			20.2	+0.33	
			22.6	−0.34	
$(+)_{589}[Co(NO_2)_2(CO_3)(en)]^-$	20.7	204	20.4	+2.66	R
$(+)_{589}[Co(NO_2)_2(en)(H_2O)_2]^+$	21.4	170	18.4	+0.05	R
			22.4	−0.27	
$(+)_{589}[Co(NO_2)_2(ox)(en)]^-$	21.1	177	21.6	+1.83	R

peak of negative sign is ignored. Thus the carbonato complex is assigned to the R configuration. The $(+)_{589}[Co(CN)_2(NH_3)_2(H_2O)_2]^+$ species derived from the R-carbonato complex should be R because of the retention of the configuration. The CD pattern of $(-)_{589}[Co(CN)_2(ox)(NH_3)_2]^-$ is almost reverse to that of R-$[Co(CN)_2(CO_3)(NH_3)_2]^-$, and hence the S configuration is assigned to the oxalato complex. As to the remaining $(+)_{589}[Co(CN)_2(ox)(en)]^-$ complex, the CD signs are identical with those of S-$[Co(CN)_2(mal)(NH_3)_2]^-$ and opposite to those of the R-$[Co(CN)_2(CO_3)(en)]^-$ complex. Consequently, this oxalato complex also has S configuration.

When a bidentate ligand is replaced by the corresponding two unidentates, or an O,O-chelate (CO_3^{2-}, ox^{2-}, or mal^{2-}) is replaced by another one, the sign of the peak on the higher frequency side of the T_{1g} peak never changes for a fixed absolute configuration; for the $[Co(CN)_2(O,O)(N)_2]$ complexes the (+) sign corresponds to S. while for the $[Co(NO_2)_2(O,O)(N)_2]$ complexes, the (+) sign corresponds to R with two exceptions of the diaquadinitro complexes.

During 1917~1937 several schools [41~44] intended to determine the geometrical structure of Erdmann's salt, $NH_4[Co(NO_2)_4(NH_3)_2]$. Their conclusions were contradictory regarding whether an oxalate derivative could be resolved into enantiomers or not. Later, X-ray studies showed that the two ammonia groups were in the *trans* positions in the silver [45,46], potassium [47] and ammonium [48] salts. On the other hand, it was claimed that in aqueous solution of Erdmann's salt, the *trans* and *cis* isomers are in equilibrium [49]. This problem remained unsolved until the Shibata group [37] prepared all three isomers of $[Co(NO_2)_2(ox)(NH_3)_2]^-$; the *cis,cis*-isomer and the *trans*(NO_2)-isomer were obtained from *cis,cis*-$[Co(ox)(H_2O)_2(NH_3)_2]^+$ and KNO_2, the *trans*(NH_3)-isomer from $[Co(NO_2)_4(NH_3)_2]^-$ and oxalic acid.

Based on the CD spectra of R(+)-*cis,cis,cis*-[Co(CN)$_2$(NH$_3$)$_2$(H$_2$O)$_2$]$^+$ and S(+)-*cis,cis,cis*-[Co(NO$_2$)$_2$(NH$_3$)$_2$(H$_2$O)$_2$]$^+$, Mason [50] investigated the optical activity theoretically, employing a third-order and a fourth-order ligand-polarization model; he concluded that both ligand-polarization mechanisms contribute significantly, but not exclusively, to the d-electron optical activity of chiral unidentate complexes of the *cis,cis,cis*-[Coa$_2$b$_2$c$_2$] type.

5.3.2 *fac*(A)-[Co(A$_3$)(BC)d], *fac*(A)-[Co(A$_3$)bcd] and *fac*(a)-[Co(a)$_3$BCd]

The ligand, 1,1,1-tris(aminomethyl)ethane (tame) is a typical tripod ligand and coordinates facially to a cobalt(III) ion. The hexaniobate ion, Nb$_6$O$_{19}^{8-}$ (Fig. 5.13), is known to function as a bulky terdentate ligand [51,52]. Shimura et al. [53,54] prepared a few complexes of the *fac*(A)-[Co(A$_3$)(BC)d] type, where A$_3$ represents a terdentate ligand.

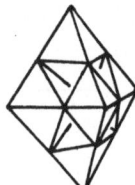

Fig. 5.13. Nb$_6$O$_{19}^{8-}$ ion

For the preparation of the complex, *cis*-K[Co(CO$_3$)$_2$(NH$_3$)$_2$] · H$_2$O was allowed to react with tame · 3 HCl [55], neutralized in an aqueous solution to produce [Co(CO$_3$)(NH$_3$)(tame)]$^+$, which was isolated as the chloride 1.5-hydrate by ion-exchange chromatography (Dowex 50W-X8 in Na$^+$ form). The aqueous solution of the intermediate reacted with glycine to form the desired complex, which was isolated as [Co(gly)MNH$_3$)(tame)]Cl$_2$ by chromatography (SP-Sephadex C-25 in Na$^+$ form). This product was resolved with K$_2$[Sb$_2$(L-tart)$_2$] · 3 H$_2$O. The resolving agent was removed from the less soluble diastereoisomer on a column of anion-exchange resin in Cl$^-$ form to obtain (−)$_{589}$[Co(gly)(NH$_3$)(tame)] Cl$_2$. Related complexes containing an optically active amino acid instead of glycine were also prepared and chromatographically separated into diastereoisomeric pairs.

The CD spectral data of the (−)$_{589}$-glycinato complex is cited in Table 5.6. The so-called additivity rule of the configurational and vicinal contributions permitted to find that the configurational curves for the diastereomeric complexes are similar to the CD curve for the enantiomeric glycinato complex.

As a reagent for the terdentate ligand, sodium hydrogenhexaniobate Na$_7$HNb$_6$O$_{19}$ · 15 H$_2$O was used [54]. The action of a neutralized solution of hexaniobate on aqueous *cis*-K[Co(CO$_3$)$_2$(NH$_3$)$_2$] · H$_2$O resulted in the isolation of K$_7$[Co(Nb$_6$O$_{19}$)(CO$_3$)-(NH$_3$)]. Using this, [Co(Nb$_6$O$_{19}$)(gly)(NH$_3$)]$^{6-}$ and [Co(Nb$_6$O$_{19}$)(gly)(H$_2$O)]$^{6-}$ were isolated as lithium salts. Optical resolutions of these complexes were carried out with (+)$_{589}$[Co(en)$_3$] Br$_3$. Related complexes containing an optically active amino acid instead of glycine were again prepared in a similar way and chromatographed into diastereoisomers.

Table 5.6. Absorption and CD Data of the tame- and Nb_6O_{19}-Containing Complexes

(T_{1g} Band)

Complex	Absorption		CD	
	$\tilde{v}/10^3$ cm^{-1}	log ε	$\tilde{v}/10^3$ cm^{-1}	Δε
$(-)_{589}[Co(gly)(NH_3)(tame)]^{2-}$	20.60	1.92	18.40	+0.016
			20.50	−0.44
$(+)_{546}^{CD}[Co(Nb_6O_{19})(gly)(NH_3)]^{6-}$	17.0	1.89	16.4	−0.21
			18.3	+0.61
$(+)_{546}^{CD}[Co(Nb_6O_{19})(gly)(H_2O)]^{6-}$	16.5	1.87	14.6	+0.20
			16.0	−0.26
			17.9	+0.08

The CD spectral data of the less soluble complexes are cited in Table 5.6. In contrast to the tame complexes, the CD spectra of the hexaniobato complexes deviate considerably from the additivity rule.

Since the 1,4,7-triazacyclononane ligand, $NHCH_2CH_2NHCH_2CH_2NHCH_2CH_2$ (tacn), coordinates facially in an octahedral complex, it serves to prepare the $fac(A)$-$[Co(A_3)(BC) d]$ complexes. Using this ligand, glycinate and a variety of unidentates, various complexes were synthesized [56].

The trihydrochloride of the terdentate ligand was prepared by the method of Richman and Atkins [57]. Equimolar amounts of $mer(N)trans(NH_3)$-$[Co(CO_3)(gly)-(NH_3)_2]$ and tacn · 3 HCl were dissolved in water, acidified to pH 2 with aqueous $HClO_4$ and then adjusted to pH 9 with aqueous KOH, whereupon the solution turned from violet to red, indicating the formation of $[Co(gly)(tacn)(H_2O)]^{2+}$. The complex was purified by ion-exchange chromatography (on SP-Sephadex C-25 in Na$^+$ form with 0.1 mol/dm^3 NaClO$_4$). Finally, from an ethanolic solution of the concentrated eluate, red crystals of $[Co(gly)(tacn)(H_2O)]$ $(ClO_4)_2$ were obtained.

From this intermediate aqua complex, the following compounds were produced:

$[Co(NO_2)(gly)(tacn)]$ Cl · H$_2$O	(yellow),
$[Co(gly)(NH_3)(tacn)]$ I$_2$ · H$_2$O	(orange),
$[Co(NCS)(gly)(tacn)]$ Br	(red),
$[Co(N_3)(gly)(tacn)]$ I · H$_2$O	(red-violet),
$[CoCl(gly)(tacn)]$ Cl · 1.5 H$_2$O	(violet),
$[CoI(gly)(tacn)]$ I	(green) .

The reaction of the aqua complex and KCN in aqueous solution gave no cyano complex, but unidentified materials. On the other hand, KCN substituted the coordinated chloride ion in $[CoCl(gly)(tacn)]NO_3$ in DMSO by the cyanide ion. Yellow crystals of $[Co(CN)(gly)(tacn)]Br$ were obtained after ion-exchange purification (Scheme 3).

Optical resolutions of the cyano, nitro and ammine complexes were carried out by column chromatography of SP-Sephadex C-25 in the Na$^+$ form. The two enantiomers were separated with 0.05 mol/dm^3 Na$_2[Sb_2(L-tart)_2]$ · 2 H$_2$O as eluent. The iso-

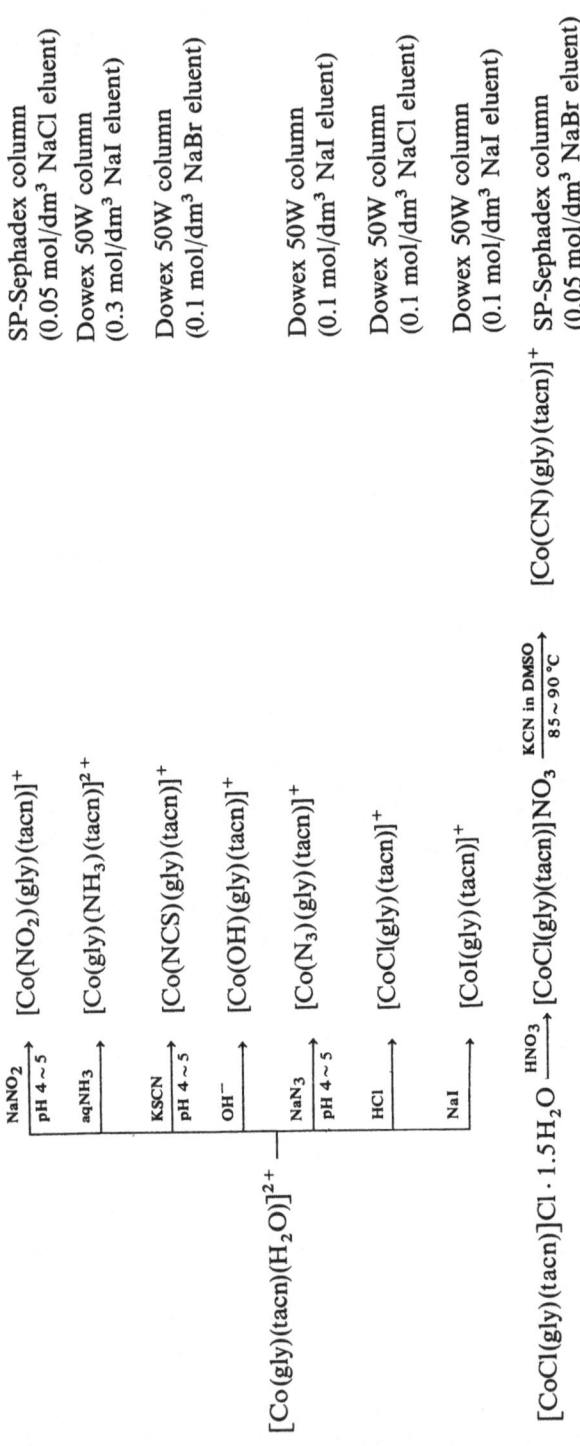

Scheme 3. Preparation Pathways and Purification Procedures

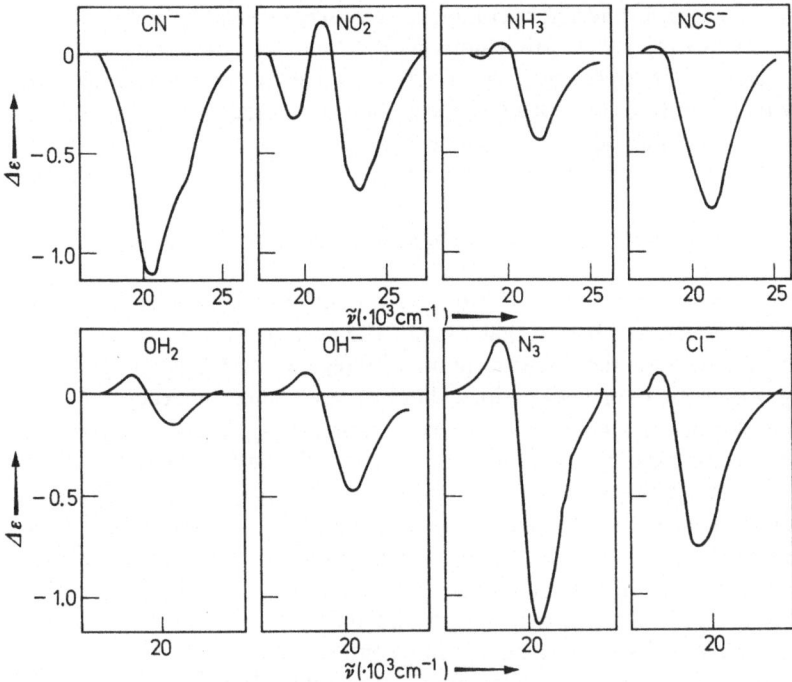

Fig. 5.14. CD spectra for earlier eluted enantiomers

Table 5.7. Absorption and CD Data of the tacn-Containing Complexes

$(T_{1g}$ Band, $\tilde{v}/10^3$ cm$^{-1})$

Complex	Absorption	CD	
	$\tilde{v}$ (log ε)	$\tilde{v}$	($\Delta\varepsilon$)
$(-)_{589}$[Co(CN)(gly)(tacn)]Cl	21.9 (1.91)	20.7	(-1.05)
		ca. 23.0	$(-0.5\ sh)$
$(-)_{589}$[Co(NO$_2$)(gly)(tacn)]Cl	21.8 (2.12)	19.8	(-0.32)
		21.3	$(+0.15)$
		23.6	(-0.66)
$(-)_{589}$[Co(gly)(NH$_3$)(tacn)]I$_2$	20.8 (1.96)	18.6	(-0.02)
		19.8	$(+0.06)$
		22.0	(-0.42)
$(-)_{589}$[Co(NCS)(gly)(tacn)]Br	20.3 (2.31)	17.8	$(+0.04)$
		21.2	(-0.73)
$(-)_{589}$[Co(gly)(OH$_2$)(tacn)](ClO$_4$)$_2$	20.0 (1.88)	18.2	$(+0.10)$
		20.6	(-0.14)
$(-)_{589}$[Co(OH)(gly)(tacn)]$^+$	19.7 (1.91)	17.6	$(+0.11)$
		20.6	(-0.45)
$(+)_{589}$[Co(N$_3$)(gly)(tacn)]I	19.4 (2.32)	18.4	$(+0.27)$
		20.8	(-1.10)
$(-)_{589}$[Co(Cl)(gly)(tacn)]Cl	18.6 (2.04)	16.9	$(+0.10)$
		19.6	(-0.71)
[Co(I)(gly)(tacn)]I	17.2 (2.20)		

thiocyanato, aqua, azido and chloro complexes were separated completely on columns of Dowex 50W-X8 resin (Na^+) by eluting with 0.3 mol/dm³ $Na_2[Sb_2(L-tart)_2]$. It is worth noting that ion-exchange resin is very effective for such a complex which is only partially resolved on Sephadex. Resolution of the iodo complex was unsuccessful since the ligating iodide is too labile.

All the CD spectra of the resolved isomers are mirror images of the enantiomeric counterparts. The CD spectra, in the T_{1g} region, for the isomers obtained from the earlier eluates are illustrated in Fig. 5.14. The absorption and CD data are summarized in Table 5.7.

A common characteristic of all CD spectra for the enantiomeric isomers, except for the CN^--containing isomer, is that a major CD peak of negative sign is observed at the higher frequency in the first absorption band region. The CD curve of the CN^--containing isomer shows one peak with a shoulder at a higher frequency.

These optically active complexes were derived from $(-)_{589}[Co(gly)(tacn)(H_2O)]^{2+}$. The derivations for the nitro, isothiocyanato, azido and chloro complexes were achieved with success. On the other hand, the oxidation of $(-)_{589}[Co(NCS)(gly)(tacn)]^+$ by aqueous H_2O_2 at pH $ca.$ 3 produced $(-)_{589}$-ammine, $(-)_{589}$-cyano and $(-)_{589}$-aqua complexes, all being separated chromatographically. In similar oxidative degradation reactions, Gillard and Maskil [58] found that $(-)-[Co(NCS)_2(en)_2]^+$ was converted into $(-)-[Co(NH_3)_2(en)_2]^{3+}$ and also to $(-)-[Co(CN)_2(en)_2]^+$, and that these three complexes have the same absolute configuration.

The X-ray analysis of $(-)_{589}[Co(gly)(NH_3)(tacn)] I_2 \cdot H_2O$ [59] has shown the absolute configuration as in Fig. 5.15. From the same sign in the major CD peaks, the earlier eluted enantiomers were assumed to have the same arrangement of the glycinate and an unidentate (NO_2^-, NH_3, NCS^-, H_2O, OH^-, N_3^-, or Cl^-). In addition, the derivations of $(-)_{589}$-NH_3 and $(-)_{589}$-CN isomers from $(-)_{589}$-NCS complex proved that these three have the same absolute configuration.

For the present chiral complexes, three sources of d-electron optical activity can be considered; herical distribution of a glycinate ring and a $-NH-CH_2-CH_2-NH-$ chelate ring of tacn, the conformation of the three chelate rings of tacn, and the arrangement of a glycinate ring and a unidentate ligand. However, the first and second sources have little effect on the solution CD spectrum: The spectra of $(-)_{589}[Co(gly)(NH_3)(tacn)]^{2+}$ and $(-)_{589}[Co(gly)(NH_3)(tame)]^{2+}$ resemble each other well in shape and peak intensity. The $\Delta\varepsilon$ value for the $(-)_{589}$-tacn complex, -0.42, corresponds well with that for the $(-)_{589}$-tame complex, -0.44.

The CD spectra of bis[(R)-2-methyl-1,4,7-triazacyclononane]cobalt(III), $[Co(R-metacn)_2]^{3+}$, were studied by Mason and Peacock [60]. For this complex, five

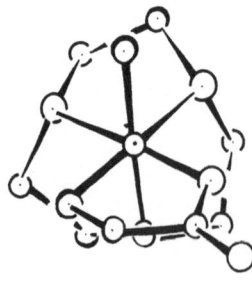

Fig. 5.15. Absolute configuration of $(-)_{589}[Co(gly)(NH_3)(tacn)]^{2+}$

isomers of nine possible geometrical isomers were isolated by chromatography on SP-Sephadex [61]. Each CD spctrum exhibited a very strong positive peak in the first absorption band region ($\Delta\varepsilon = +4.35 \sim +4.72$), due to chiral puckering, with the λ-conformation, of each of the six chelate rings [60,61]. The crystal structure was determined with a mixture of the three isomers [62]. In this connection the Shibata group [56] found that when CD measurements were carried out for aqueous $[Co(tacn)_2]^{3+}$ in the presence of $[Sb_2(\text{L-tart})_2]^{2-}$, a positive peak in the first absorption band region indicated prefered formation of a conformation (probably δ) of the chelate rings. This fact indicates that the tacn chelate rings have no fixed conformation in a solution, and hence a contribution from the conformation of the chelate rings to the solution CD spectra is negligible, though the crystal structure of the complex reveals either $\lambda\lambda\lambda$ or $\delta\delta\delta$ ring conformation.

As a complex of fac(A)-$[Co(A_3)$ bcd]-type, amminebromocyano(1,4,7-triazacyclo-nonane)cobalt(III) chloride, $[Co(Br)(CN)(NH_3)(tacn)]$ Cl was prepared and resolved into the enantiomers [63]. The ligand substitutions occurred in two steps:

i) $[Co(CN)(SO_3)(NH_3)_4]$ [64] $\xrightarrow[\text{50 °C, pH 8} \sim 9]{\text{tacn} \cdot 3\,HCl}$ $\xrightarrow[\text{H}_2\text{O eluent}]{\text{SP-Sephadex column}}$

$[Co(CN)(SO_3)(NH_3)(tacn)] \cdot 0.5\,H_2O$

ii) $[Co(CN)(SO_3)(NH_3)(tacn)]$ $\xrightarrow[\text{50 °C}]{\text{HBr}}$ $\xrightarrow{\text{EtOH-(Et)}_2\text{O}}$ solid

$\xrightarrow[\text{(0.05 mol/dm}^3\text{ NaCl eluent)}]{\text{SP-Sephadex column}}$ $[Co(Br)(CN)(NH_3)(tacn)]^+$

Optical resolution was achieved by chromatography on SP-Sephadex C-25 in Na$^+$ form with $K_2[Sb_2(\text{L-tart})_2]$.

A novel complex, triammineglycinatonitrocobalt(III), of fac(a)-$[Co(a)_3(BC)d]$ type was prepared [65] as follows:

$[Co(NO_2)(NH_3)_5]^{2+}$ $\xrightarrow[\substack{\text{charcoal,} \\ \text{room temp.}}]{\text{Hgly, Na}_2\text{CO}_3}$ $\xrightarrow[\text{chromatography}]{\text{Ion/exchange}}$

fac(NH$_3$)-$[Co(NO_2)(gly)(NH_3)_3]$ Cl

During the chromatographic purification, two other isomers, mer(NH$_3$)$trans$(NO$_2$, O) and mer(NH$_3$)$trans$(NO$_2$, N), were also separated. The total resolution of the fac(NH$_3$) complex was attained by elution with a $K_2[Sb_2(\text{L-tart})_2]$ solution. The absorption and CD spectra of an enantiomer of the fac(NH$_3$) complex are illustrated in Fig. 5.16, together with those of related complex.

5.4 Chiral Complexes of the $[Co(O)_4(N)_2]^-$ and $[Co(O)_2(N)_4]^+$ Types

5.4.1 Cis(N)-$[Co(O,O)_2(N)_2]^-$ Complexes

Douglas et al. [66] took up bis(dicarboxylato)ethylenediamine complexes such as $[Co(ox)_2(en)]^-$ and $[Co(mal)_2(en)]^-$ as crystal field models of the ethylenediamine-tetraacetatocobaltate(III) ion and compared the absorption and CD spectra; the

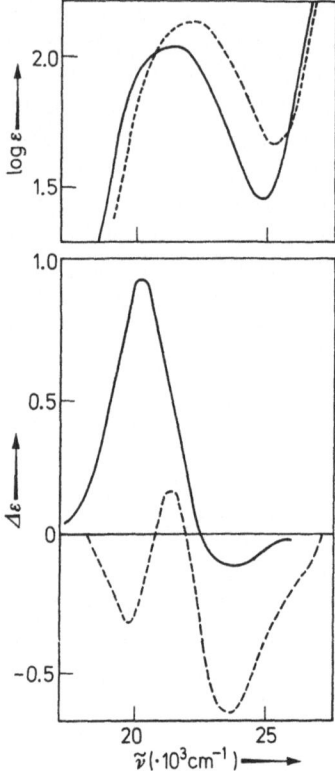

Fig. 5.16. Absorption and CD spectra of *fac*(NH₃)-[Co(NO₂)(gly)(NH₃)₃]⁺ (————), and [Co(NO₂)(gly)(tacn)]⁺ (– – – –) (from Ref. [65])

CD curve for (—)₅₄₆[Co(edta)]⁻ whose absolute configuration had been assigned to Λ by comparison of the ORD curve with that of Δ(+)₅₄₆[Co(—)pdta]⁻ (pdta = 1,2-propylenediaminetetraacetate) [67] showed a (+, —) pattern under the first absorption band (T₁g) and (+,+) pattern under the second absorption band (T₂g). The CD curve for (+)₅₄₆[Co(ox)₂(en)]⁻ showed two peaks of the same sign (+,+) in the T₁g region and three peaks of alternating signs (—,+,—) in the T₂g region. The CD curve for (—)₅₄₆[Co(mal)₂(en)]⁻ exhibited three well defined peaks of alternating signs (+,—,+) in both the T₁g and T₂g regions (Table 5.8).

From the sign of the component at the lowest frequency under the T₁g band, (—)₅₄₆[Co(ox)₂(en)]⁻ was assigned to Δ and (+)₅₄₆[Co(mal)₂(en)]⁻, to Λ, the latter configuration being supported by X-ray studies [68,69]. The CD spectra of the corresponding complexes containing trimethylenediamine instead of ethylenediamine were also studied by Brennan et al. [70]; as far as the T₁g region was concerned, the CD curves for (+)₅₄₆[Co(ox)₂(tn)]⁻ and (—)₅₄₆[Co(mal)₂(tn)]⁻ showed the same patterns as those for the corresponding ethylenediamine complexes (Table 5.8).

These results show that in the [Co(O,O)₂(diamine)]⁻-type complexes the variation of O,O-donor bidentates causes a change in the CD pattern observed. Thus, it seems worthwhile to examine the CD spectra for a series of *cis*(N)-[Co(O,O)₂(N)₂]⁻-type complexes, where (N)₂ represents a diamine or two unidentate N-donor ligands. Shibata et al. [71,72] prepared such complexes based on *cis*-substitution of the ligating carbonate ion in a starting complex. The pathway is given in Scheme 4.

Table 5.8. Absorption and CD Data for $[Co(O)_4(N)_2]^-$ Complexes

$(T_{1g}$ Band, $\tilde{v}/10^3$ cm^{-1})

Complex	Absorption		CD		Absolute configu- ration
	$\tilde{v}$	ε	$\tilde{v}$	$\Delta\varepsilon$	
$K(-)_{546}[Co(edta)] \cdot 2\,H_2O$	18.60	347	17.30	+1.50	Λ
			19.83	−0.76	
			21.75	−0.26	
$Na(+)_{546}[Co(ox)_2(en)] \cdot H_2O$	18.50	109	17.20	−2.27	Δ
			18.60	+0.39	
			20.20	−0.82	
$K(-)_{546}[Co(mal)_2(en)] \cdot 2\,H_2O$	18.50	95	16.95	+3.09	Λ
			18.55	−2.96	
			20.55	+1.06	
$K(+)_{546}[Co(ox)_2(tn)] \cdot 2\,H_2O$	18.0	101	17.2	−1.61	Δ
			19.3	+0.67	
			23.8	−0.17	
$K(-)_{546}[Co(mal)_2(tn)] \cdot 2\,H_2O$	18.1	78	16.6	+2.09	Λ
			18.2	−2.10	
			20.3	+0.68	

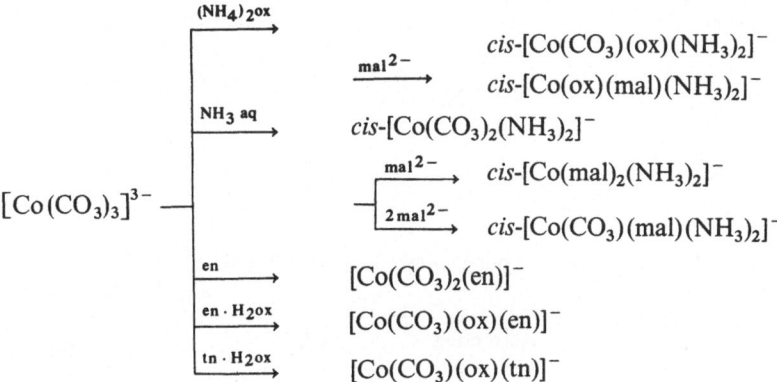

Scheme 4. The Preparative Pathway for $[Co(O,O)_2(diamine)]^-$-type Complexes

As an example, the preparation of $Na[Co(ox)(mal)(en)] \cdot H_2O$ is briefly described. An aqueous solution of $K[Co(CO_3)(ox)(en)] \cdot H_2O$ (0.1 mol) was mixed with a solution of malonic acid (0.15 mol) and then stirred at 40 °C for 2 h. The filtrate was chromatographed on a column of anion-exchange resin (Dowex 1-X8, Cl$^-$ form). By elution with 0.1 mol/dm^3 NaCl, three bands descended. From the second fraction, the desired complex was obtained. The products from the first and third were $Na[Co(mal)_2(en)]$ and $Na[Co(ox)_2(en)]$, respectively.

Two kinds of resolving agents, $(-)_{589}[Co(ox)(en)_2] (C_2H_3O_2)$ and $(-)_{589}[Co-(NO_2)_2(en)_2] (C_2H_3O_2)$, were successfully used for the optical resolution of the complexes except $[Co(CO_3)_2(NH_3)_2]^-$. However, an optically active lithium salt of this dicarbonato complex was isolated by an unusual way [73]. The potassium salt, K[Co-$(CO_3)_2(NH_3)_2] \cdot H_2O$, was converted to the lithium salt by ion-exchange chromatography. An aqueous solution of the lithium salt (0.02 mol) was added to dissolved $(-)_{589}[Co(ox)(en)_2] (C_2H_3O_2)$ (0.01 mol). After ammonium carbonate had been added to prevent decomposition of the dicarbonato complex, the solution was concentrated under reduced pressure and then kept cool, whereupon Li$(-)_{589}$[Co-$(CO_3)_3(NH_3)_2]$ separated out with ca. 70% yield. The isolated complex was optically labile so that its optical rotation decreased, in aqueous solution, with a half life of ca. 3 min. at room temperature.

In so far documented asymmetric syntheses of metal complexes, optical lability

Table 5.9. Spectral Data and Absolute Configurations for $[Co(O,O)_2(N)_2]$-type Complexes

(T_{1g} Band, $\tilde{v}/10^3$ cm^{-1})

Complex	Absorption	CD		Absolute configuration
	$\tilde{v}$ (log ε)	$\tilde{v}$	($\Delta\varepsilon$)	
$(+)_{589}[Co(CO_3)_2(en)]^-$	17.5 (2.17)	17.6	(−1.91)	Δ
$(-)_{589}[Co(CO_3)(ox)(en)]^-$	18.1 (2.16)	17.2	(+3.43)	Λ
$(+)_{589}[Co(ox)(en)(H_2O)_2]^+$	18.1 (1.94)	17.3	(−0.47)	Λ
		19.7	(+0.99)	
$(-)_{589}[Co(ox)(mal)(en)]^-$	18.5 (2.04)	17.0	(+2.80)	Λ
		18.0	(−1.29)	
		20.6	(+0.71)	
$(+)_{589}[Co(CO_3)_2(py)_2]^-$	17.7 (2.24)	18.4	(+2.94)	Λ
$(+)_{589}[Co(ox)_2(py)_2]^-$	18.2 (2.06)	18.2	(+1.53)	Λ
$(+)_{589}[Co(CO_3)(ox)(tn)]^-$	17.8 (2.08)	18.3	(+1.27)	Λ

Table 5.10. Spectral Data and Absolute Configurations for $[Co(O,O)_2(NH_3)_2]$-type Complexes

(T_{1g} Band, $\tilde{v}/10^3$ cm^{-1})

Complex	Absorption	CD		Absolute configuration
	$\tilde{v}$ (log ε)	$\tilde{v}$	($\Delta\varepsilon$)	
$(-)_{589}[Co(CO_3)_2(NH_3)_2]^-$	17.3 (2.13)	17.8	(−2.32)	Δ
$(-)_{589}[Co(CO_3)(ox)(NH_3)_2]^-$	17.7 (2.09)	17.3	(−1.74)	Δ
$(+)_{589}[Co(ox)_2(NH_3)_2]^-$	18.0 (2.04)	17.3	(−1.52)	Δ
		20.2	(−0.46)	
$(+)_{589}[Co(ox)(mal)(NH_3)_2]^-$	18.0 (1.95)	16.7	(−2.25)	Δ
		18.5	(+2.06)	
		20.9	(−0.31)	
$(+)_{589}[Co(mal)_2(NH_3)_2]^-$	18.0 (1.92)	16.5	(−1.74)	Δ
		18.3	(+2.52)	
		20.6	(−0.41)	
$(+)_{589}[Co(CO_3)(mal)(NH_3)_2]^-$	17.8 (2.00)	16.5	(−1.21)	Δ
		18.4	(+0.81)	
$(-)_{589}[Co(mal)(NH_3)_2(H_2O)_2]^+$	18.3 (1.72)	17.0	(+0.43)	S
		19.4	(−0.49)	

and formation of the less soluble diastereoisomer with a resolving agent have been considered to be essential for the syntheses [74]. On the other hand, the essential conditions for the isolation of one enantiomeric salt are the poorer solubility of an enantiomeric salt than that of the racemic salt in an aqueous solution containing an optically active complex cation, the different solubilities between the enantiomeric salts, and the rapid racemization of the enantiomeric complex species.

The absorption and CD spectral data are collected in Tables 5.9 and 5.10. The CD spectra could be grouped into four types of patterns in the T_{1g} region; 1) only one peak, 2) two peaks of opposite signs, 3) two peaks of the same sign, and 4) three peaks of alternating sign. The active dicarbonato complexes exhibit a symmetrical peak of type 1. The active carbonatooxalato complexes also show only one peak, which is unsymmetrical. The $(+)_{589}[Co(CO_3)(mal)(NH_3)_2]^-$ complex exhibits the pattern of type 2. The $(-)_{546}[Co(ox)_2(en)]^-$ complex has the pattern of type 3. The active oxalatomalonato and bis(malonato) complexes show the pattern of type 4.

The absolute configurations of the complexes were determined by comparison of the patterns with those for the optically active $[Co(mal)_2(en)]^-$, $[Co(ox)_2(en)]^-$ and $[Co(edta)]^-$ complexes (Tables 5.9 and 5.10). The CD patterns in the T_{1g} region for the

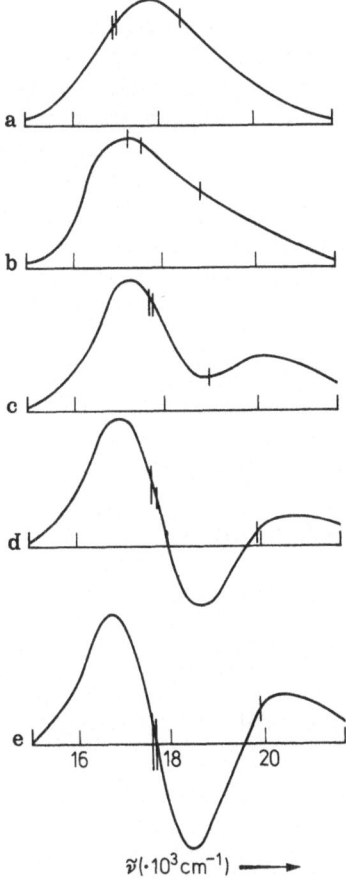

a

b

c

d

e

16   18   20

$\tilde{\nu}(\cdot 10^3 cm^{-1})$ ⟶

Fig. 5.17a–e. Variation of CD spectra in the $[Co(O,O)_2(en)]^-$-type complexes of Λ configuration. **a** $(-)_{589}[Co(CO_3)_2(en)]^-$, **b** $(-)_{589}[Co(CO_3)(ox)(en)]^-$, **c** $(-)_{546}[Co(ox)_2(en)]^-$, **d** $(-)_{589}[Co(ox)(mal)(en)]^-$, **e** $(-)_{546}[Co(mal)_2(en)]^-$ (from Ref. [71])

complexes having Λ configuration are collected in Fig. 5.17, where the peak intensity is arbitary and the short vertical lines on each CD curve indicate the band positions calculated by means of Yamatera's prediction.

Figure 5.17 clearly shows regular changes in the patterns. In order to understand these regular changes, the workers attempted to analyze the patterns into Gaussian curves; the Gaussian analysis showed that every pattern in the T_{1g}-band region results from the mutual cancellation of a $(+)$ component at a lower frequency and a $(-)$ component at a higher frequency, and that the resultant CD sign at the higher frequency is altered by mutual cancellation, while the sign at the lower frequency remains unchanged. On this basis the absolute configurations of the $[Co(O,O)_2(N)_2]^-$-type complexes can be assigned from the sign of the lowest-frequency peak; the $(+)$ sign refers to the Λ isomer and the $(-)$ sign to the Δ isomer. It should be noted that for the $[Co(O,O)(N)_2(H_2O)_2]^+$-type complexes, the $(-)$ sign relates to the Λ configuration of the corresponding $[Co(O,O)(CO_3)(N)_2]^-$ complex.

5.4.2 Chiral Complexes of $[Co(N)_4(O)_2]^+$ Type

In their CD spectra, Λ-$[Co(en)_3]^{3+}$ and Λ-$[Co(tn)_3]^{3+}$ have opposite signs under the T_{1g} band [75], suggesting causality in CD sign and ring-member of ligating diamine. Furthermore, as just mentioned, a chiral complex containing a CO_3^{2-} ligand and the corresponding diaqua complex show marked difference in CD patterns in the T_{1g} band region. However, when the acidified solution containing such a diaqua complex is alkalized with an excess of bicarbonate, the CD spectrum of the solution returns to that of the parent carbonato complex, indicating the retention of the absolute configuration through hydrolysis. In relation to such drastic changes of CD patterns, Shibata et al. [72,76] designed some carbonato complexes of the $[Co(CO_3)(N)_4]^+$ type, where $(N)_4$ represents $2\,NH_3+en$, $en+2py$, $2\,NH_3+bpy$, $2\,NH_3+2\,py$, or $2\,C_2H_5NH_2+2\,py$, in order to investigate CD spectral changes between those carbonato complexes and the acid-hydrolyzed diaqua complex species. The preparative pathways are illustrated by Scheme 5.

$$[Co(CO_3)_2(en)]^- \xrightarrow[\substack{\text{chromato-}\\\text{graphy*}}]{\substack{\text{conc. } NH_3\\\text{or py}}} \quad \begin{array}{l} cis\text{-}[Co(CO_3)(NH_3)_2(en)]^+ \\ \text{or} \\ cis\text{-}[Co(CO_3)(en)(py)_2]^+ \end{array}$$

$$cis\text{-}[Co(CO_3)_2(NH_3)_2]^- \xrightarrow[\substack{\text{chromato-}\\\text{graphy*}}]{\substack{\text{bpy}\\\text{or py}}} \quad \begin{array}{l} cis\text{-}[Co(CO_3)(NH_3)_2(bpy)]^+ \\ \text{or} \\ cis,cis\text{-}[Co(CO_3)(NH_3)_2(py)_2]^+ \end{array}$$

$$cis\text{-}[Co(CO_3)_2(NH_2C_2H_5)_2]^- \xrightarrow[\substack{\text{chromato-}\\\text{graphy*}}]{\text{py}} \quad cis,cis\text{-}[Co(CO_3)(NH_2C_2H_5)_2(py)_2]^+$$

* Dowex 50W-X8, Na$^+$ form, 0.2 or 0.3 mol/dm^3 NaCl eluent

Scheme 5. Preparative Pathways for $[Co(CO_3)(N)_4]^+$ Complexes

The complex, cis-[Co(CO$_3$)(NH$_3$)$_2$(en)] Cl · H$_2$O, was first prepared by Bailar and Peppard [33], who used the reaction of cis(NH$_3$)trans(Cl)-[CoCl$_2$(NH$_3$)$_2$(en)]$^+$ and Ag$_2$CO$_3$, in which the dichloro complex was rearranged to produce the carbonato complex. Shibata et al. [72] prepared the same compound by cis-substitution of [Co(CO$_3$)$_2$(en)]$^-$ with NH$_3$ ligands, and resolved the complex with (−)$_{546}$[Co(ox)$_2$-(en)]$^-$; the complex could also be resolved with [Co(SO$_3$)$_2$(NH$_3$)$_2$(R-pn)]$^-$ [33]. The resolutions of [Co(CO$_3$)(NH$_3$)$_2$(py)$_2$]$^+$ and [Co(CO$_3$)(en)(py)$_2$]$^+$ were achieved with the same agent. The other complexes, [Co(CO$_3$)(NH$_3$)$_2$(bpy)]$^+$ and [Co(ox)-(NH$_3$)$_2$(py)$_2$]$^+$, were resolved with (+)$_{546}$[Co(edta)]$^-$; [Co(CO$_3$)(NH$_2$C$_2$H$_5$)$_2$(py)$_2$]$^+$ was resolved with (−)$_{546}$[Co(ox)$_2$(en)]$^-$.

The absorption spectral and CD spectral data are summarized in Table 5.11; the data on the diaqua complexes were obtained with the complex derived from the corresponding optically active carbonato complexes by acid hydrolysis.

With respect to (+)$_{589}$[Co(CO$_3$)(NH$_3$)$_2$(en)]$^+$, (+)$_{546}$[Co(CO$_3$)(en)$_2$]$^+$ and (+)$_{589}$-[Co(CO$_3$)(NH$_3$)$_2$(bpy)]$^+$, only one (+) CD peak was observed in the T$_{1g}$ region, the corresponding diaqua complex exhibiting two CD peaks of (−,+) signs. The (−)$_{589}$[Co(CO$_3$)(NH$_3$)$_2$(py)$_2$]$^+$ complex exhibits a (−,+) pattern, and the cor-

Table 5.11. Absorption and CD Spectral Data for Chiral [Co(O)$_2$(N)$_4$]$^+$ Complexes

($\tilde{v}/10^3$ cm^{-1})

Complex	Absorption		CD (T$_{1g}$ band)		Transition component in the holohedrized symmetry D$_{4h}$	Absolute configuration
	$\tilde{v}$	log ε	$\tilde{v}$	Δε		
(+)$_{546}$[Co(CO$_3$)(en)$_2$]$^+$ [a]	19.4	2.16	18.7	+3.70	A$_{2g}$	Λ
(+)$_{546}$[Co(en)$_2$(H$_2$O)$_2$]$^{3+}$ [a]	20.2	1.92	17.9	−0.3	A$_{2g}$	Λ
			20.6	+1.05	E$_g$	
(+)$_{589}$[Co(CO$_3$)(NH$_3$)$_2$(en)]$^+$	19.4	2.08	18.8	+1.53	A$_{2g}$	Λ
(+)$_{589}$[Co(NH$_3$)$_2$(en)(H$_2$O)$_2$]$^{3+}$	20.1	1.97	18.1	−0.17	A$_{2g}$	R(Λ)
			20.6	+0.22	E$_g$	
(+)$_{589}$[Co(CO$_3$)(NH$_3$)$_2$(bpy)]$^+$	19.5	2.07	19.0	+2.03	A$_{2g}$	Λ
(+)$_{589}$[Co(NH$_3$)$_2$(bpy)(H$_2$O)$_2$]$^{3+}$	20.1	1.82	18.2	−0.29	A$_{2g}$	R(Λ)
			21.1	+0.66	E$_g$	
(−)$_{589}$[Co(CO$_3$)(NH$_3$)$_2$(py)$_2$]$^+$	19.4	2.06	18.1	−0.59	A$_{2g}$	S(Δ)
			20.3	+0.92	E$_g$	
(−)$_{589}$[Co(NH$_3$)$_2$(py)$_2$(H$_2$O)$_2$]$^{3+}$	19.7	1.83	17.4	+0.15	A$_{2g}$	S(Δ)
			19.8	−0.22	E$_g$	
(+)$_{589}$[Co(CO$_3$)(en)(py)$_2$]$^+$	19.5	2.09	20.0	+1.87	E$_g$	Δ
(−)$_{589}$[Co(en)(py)$_2$(H$_2$O)$_2$]$^{3+}$	20.1	1.92	17.4	+0.59	A$_{2g}$	S(Δ)
			19.8	−0.72	E$_g$	
(+)$_{589}$[Co(CO$_3$)(NH$_2$C$_2$H$_5$)$_2$(py)$_2$]$^+$	19.2	2.08	18.1	+0.61	A$_{2g}$	R(Λ)
			20.2	−0.77	E$_g$	
(+)$_{589}$[Co(NH$_2$C$_2$H$_5$)$_2$(py)$_2$-(H$_2$O)$_2$]$^{3+}$	19.1	1.90	17.4	−0.11	A$_{2g}$	R(Λ)
			19.6	+0.24	E$_g$	
(+)$_{589}$[Co(ox)(en)$_2$]$^+$ [b]	20.0	2.01	19.2	+2.59	A$_{2g}$	Λ

a McCaffery, A. J., Mason, S. F., Norman, B. J.: J. Chem. Soc. 1965, 5094
b Jordan, W. J., Brennan, B. J., Froebe, L. R., Douglas, B. E.: Inorg. Chem. 12, 1827 (1973)

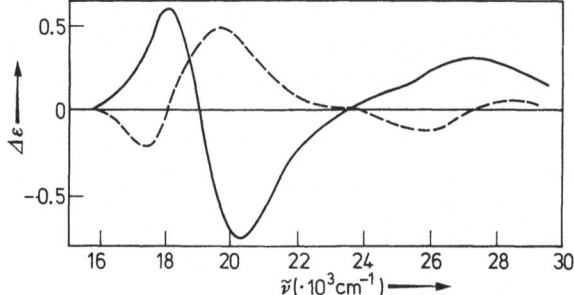

Fig. 5.18. CD spectra for $(+)_{589}[Co(CO_3)(NH_2C_2H_5)_2$-$(py)_2]^+$ (———), and $(+)_{589}[Co(NH_2C_2H_5)_2(py)_2$-$(H_2O)_2]^{3+}$ (– – – –)

responding diaqua complex exhibits the reversed $(+,-)$ pattern. The $(+)_{589}$-$[Co(CO_3)(en)(py)_2]^+$ complex exhibits a $(+)$ pattern, while the corresponding diaqua complex shows a $(+,-)$ pattern. The CD spectra of $(+)_{589}[Co(CO_3)$-$(NH_2C_2H_5)_2(py)_2]^+$ and $(+)_{589}[Co(NH_2C_2H_5)_2(py)_2(H_2O)_2]^{3+}$ are illustrated in Fig. 5.18, where reversed patterns are observed between the two spectra.

The absolute configuration of $(+)_{546}[Co(ox)(en)_2]^+$ has been determined to be Λ [77]. In $[Co(O)_2(N)_4]^+$-type complexes, the $A_{1g} \rightarrow E_g$ (in the holohedrized symmetry D_{4h}) transition lies at a higher frequency than the $A_{1g} \rightarrow A_{2g}$ transition. On the basis of this assignment of the transition components, the workers regarded the $(+)$ peak of $(+)_{546}[Co(ox)(en)_2]^+$ to be A_{2g} component. On the basis of the CD sign of this Λ-$[Co(ox)(en)_2]^+$ complex, the absolute configurations of the other complexes were determined from the CD spectra as given in Table 5.11.

5.5 References

1. Basolo, F., Ballhausen, C. J., Bjerrum, J.: Acta Chem. Scand. *9*, 810 (1955)
2. Shimura, Y., Tsuchida, R.: Bull. Chem. Soc. Jpn. *29*, 311 (1956)
3. Shimura, Y.: Bull. Chem. Soc. Jpn. *31*, 173 (1958)
4. Yamatera, H.: Bull. Chem. Soc. Jpn. *31*, 95 (1958)
5. Matsuoka, N., Hidaka, J., Shimura, Y.: Bull. Chem. Soc. Jpn. *39*, 1257 (1966)
6. Kanazawa, S., Shibata, M.: Bull. Chem. Soc. Jpn. *48*, 2424 (1971)
7. Nakai, K., Kanazawa, S., Shibata, M.: Bull. Chem. Soc. Jpn. *45*, 3544 (1972)
8. Nakashima, S., Shibata, M.: Bull. Chem. Soc. Jpn. *47*, 2069 (1974)
9. Schneider, P. W., Collman, J. P.: Inorg. Chem. 7, 2010 (1968)
10. Ida, Y., Sakai, M., Sakai, S., Shibata, M.: Bull. Chem. Soc. Jpn. *50*, 2807 (1977)
11. Jørgensen, C. K.: Acta Chem. Scand. *9*, 1362 (1955)
12. Furlani, C., Morpurgo, G., Sartori, G.: Z. Anorg. Allg. Chem. *99*, 625 (1966)
13. Emmenegger, F. P., Schwarzenbach, G.: Helv. Chim. Acta *99*, 625 (1966)
14. Nakashima, S., Shibata, M.: Bull. Chem. Soc. Jpn. *48*, 3128 (1975)
15. Ida, Y., Kobayashi, K., Shibata, M.: Chem. Lett. *1974*, 1299
16. Ida, Y., Fujinami, S., Shibata, M.: Bull. Chem. Soc. Jpn. *50*, 2665 (1977)
17. Yamada, S., Hidaka, J., Douglas, B. E.: Inorg. Chem. *10*, 2187 (1971)

18. Fujinami, S., Shibata, M., Yamatera, H.: Bull. Chem. Soc. Jpn. *51*, 1391 (1978)
19. Glerup, J., Mønsted, O., Schaeffer, C. E.: Inorg. Chem. *15*, 1399 (1976)
20. Glerup, J., Schaeffer, C. E.: Inorg. Chem. *15*, 1408 (1976)
21. Yoneda, H., Kida, S.: J. Am. Chem. Soc. *82*, 2139 (1960)
22. Ohashi, K., Fujita, J., Shimoyama, B., Saito, K.: Bull. Chem. Soc. Jpn. *41*, 2422 (1968)
23. Udavenko, V. V., Stepanenko, O. N., Erashok, B. G.: Zh. Neorg. Khim. *17*, 2690 (1972)
24. Ogino, K., Uchida, T., Nishide, T., Fujita, J., Saito, K.: Chem. Lett. *1973*, 679
25. Buckingham, D. A., Davis, C. E., Sargeson, A. M.: J. Am. Chem. Soc. *92*, 6159 (1970)
26. Nishide, T., Ogino, K., Fujita, J., Saito, K.: Bull. Chem. Soc. Jpn. *47*, 3157 (1974)
27. Okazaki, K., Shibata, M.: Bull. Chem. Soc. Jpn. *52*, 1391 (1979)
28. Miyamae, H., Saito, Y.: Acta Cryst. *B34*, 937 (1978)
29. Schaeffer, C. E.: Pure Appl. Chem. *24*, 3611 (1970); Struct. Bonding *14*, 69 (1973)
30. Hawkins, C. J.: Absolute Configuration of Metal Complex. New York, N.Y.: Wiley 1971
31. Hawkins, C. J., Niethe, J. A., Wong, C. L.: Chem. Commun. *1968*, 427
32. Hawkins, C. J., Stark, J. A., Wong, C. L.: Aust. J. Chem. *25*, 273 (1972)
33. Bailar, J. C., Jr., Peppard, D. F.: J. Am. Chem. Soc. *62*, 105 (1940)
34. Hawkins, C. J., Larsen, E., Olsen, I.: Acta Chem. Scand. *19*, 1915 (1965)
35. Cahn, R. S., Ingold, C. K., Prelog, V.: Angew. Chem. Int. (ed.): Engl. *5*, 385 (1966)
36. Enomoto, Y., Ito, T., Shibata, M.: Chem. Lett. *1974*, 423
37. Ito, T., Shibata, M.: Chem. Lett. *1975*, 375
38. Ito, T., Shibata, M.: Inorg. Chem. *16*, 108 (1977)
39. Toriumi, K., Sato, S., Saito, Y.: Acta Cryst. *B33*, 1378 (1977)
40. Shintani, H., Sato, S., Saito, Y.: Acta Cryst. *B32*, 1184 (1976)
41. Shibata, Y., Maruki, T.: J. Coll. Sci. Im. Univ. Tokyo *41*, Art. 2, 1 (1917)
42. Riesenfeld, E., Klement, R.: Z. Anorg. Chem. *124*, 1 (1922)
43. Thomas, W.: J. Chem. Soc. *123*, 617 (1923)
44. Ray, B. C.: J. Indian Chem. Soc. *14*, 440 (1937)
45. Wells, A. F.: Z. Kris. (A) *95*, 74 (1936)
46. Gilinskaya, E. A.: Ser. Fiz-Mat. i Estestven. Nauk *3*, 133 (1953)
47. Komiyama, Y.: Bull. Chem. Soc. Jpn. *29*, 300 (1953)
48. Komiyama, Y.: Bull. Chem. Soc. Jpn. *30*, 13 (1954)
49. Kuroya, H., Yamasaki, K.: Read at the Annual Meeting of Chemical Society of Japan, 1949
50. Mason, S. F.: Mol. Phys., *37*, 843 (1979)
51. Flynn, C. M., Jr., Stucky, G. D.: Inorg. Chem. *8*, 178 (1969)
52. Flynn, C. M., Jr., Stucky, G. D.: Inorg. Chem. *8*, 333 (1969)
53. Yamanari, K., Hidaka, J., Shimura, Y.: Bull. Chem. Soc. Jpn. *48*, 1653 (1975)
54. Hosokawa, Y., Hidaka, J., Shimura, Y.: Bull. Chem. Soc. Jpn. *48*, 3175 (1975)
55. Fleischer, E. B., Gebala, A. E., Lever, A., Tasker, P. A.: J. Org. Chem. *36*, 3071 (1971)
56. Shibata, M., Fujinami, S., Shimba, S.: ACS Symposium Ser., No. 119, 289 (1980)
    Shimba, S., Fujinami, S., Shibata, M.: Bull. Chem. Soc. Jpn. *53*, 2523 (1980)
57. Richman, J. E., Atkins, T. J.: J. Am. Chem. Soc. *96*, 2268 (1974)
58. Gillard, R. D.,Maskill, R.: J. Chem. Soc. (A) *1971*, 2813
59. Sato, S., Ohba, S., Shimba, S., Fujinami, S., Shibata, M., Saito, Y.: Acta Cryst. *B36*, 43 (1980)
60. Mason, S. F., Peacock, R. D.: Inorg. Chim. Acta *19*, 75 (1976)
61. Nonoyama, M.: Inorg. Chim. Acta *29*, 211 (1978)
62. Mikami, M., Kuroda, R., Konno, M., Saito, Y.: Acta Cryst. *B33*, 1485 (1977)
63. Shimba, S., Fujinami, S., Shibata, M.: Chem. Lett. *1979*, 783
64. Siebert, V. H.: Z. Anorg. Allg. Chem. *327*, 63 (1964)
65. Fujinami, S., Shibata, M.: Chem. Lett. 1981, 495
66. Douglas, B. E., Haines, R. A., Bruchmiller, J. G.: Inorg. Chem. *2*, 1194 (1963)
67. MacDermott, T. E., Sargeson, A. M.: Australian J. Chem. *16*, 334 (1963)
68. Matsumot, K., Kuroya, H.: Bull. Chem. Soc. Jpn. *44*, 3491 (1971); *45*, 1755 (1972)
69. Butler, K. R., Snow, M. R.: Chem. Comm. *1971*, 550; J. C. S. Dalton *1976*, 259
70. Brennan, B. J., Igi, K., Douglas, B. E.: J. Coord. Chem. *4*, 19 (1974)
71. Muramoto, S., Kawase, K., Shibata, M.: Bull. Chem. Soc. Jpn. *51*, 3505 (1978)
72. Tsuji, K., Fujinami, S., Shibata, M.: Bull. Chem. Soc. Jpn. *54*, 1531 (1981)

73. Muramoto, S., Shibata, M.: Chem. Lett. *1977*, 1499
74. Dwyer, F. P., Mellor, D. P. (ed.): Chelating Agents and Metal Chelates. New York and London: Academic Press 1964, Chapt. *5*, p. 195
75. Judkins, R. R., Royer, D. J.: Inorg. Chem. *13*, 945 (1974)
76. Fujinami, S., Tsuji, K., Musa, R., Shibata, M.: Bull. Chem. Soc. Jpn. *55*, 617 (1982)
77. Aoki, T., Matsumoto, K., Ooi, S., Kuroya, H.: Bull. Chem. Soc. Jpn. *46*, 159 (1973)

Author Index Volumes 101–110

The volume numbers are printed in italics

A. F. Williams

A Theoretical Approach to Inorganic Chemistry

1979. 144 figures, 17 tables. XII, 316 pages
ISBN 3-540-09073-8

Contents: Quantum Mechanics and Atomic Theory. – Simple Molecular Orbital Theory. – Structural Applications of Molecular Orbital Theory. – Electronic Spectra and Magnetic Properties of Inorganic Compounds. – Alternative Methods and Concepts. – Mechanism and Reactivity. – Descriptive Chemistry. –Physical and Spectroscopic Methods. – Appendices. – Subject Index.

This book outlines the application of simple quantum mechanics to the study of inorganic chemistry, and shows its potential for systematizing and understanding the structure, physical properties, and reactivities of inorganic compounds. The considerable strides made in inorganic chemistry in recent years necessitate the establishment of a theoretical framework if the student is to acquire a sound knowledge of the subject. A wide range of topics is covered, and the reader is encouraged to look for further extensions of the theories discussed. The book emphasizes the importance of the critical application of theory and, although it is chiefly concerned with molecular orbital theory, other approaches are discussed. This text is intended for students in the latter half of their undergraduate studies. (235 references)

Springer-Verlag
Berlin
Heidelberg
NewYork

Inorganic Chemistry Concepts

Editors: C. K. Jørgensen, M. F. Lappert, S. J. Lippard, J. L. Margrave, K. Niedenzu, H. Nöth, R. W. Parry, H. Yamatera

Volume 8
M. T. Pope

Heteropoly and Isopoly Oxometalates

1983. 71 figures, 40 tables. Approx. 240 pages
ISBN 3-540-11889-6

This book is a comprehensive survey and discussion of the chemistry, structures and applications of the extensive class of polynuclear oxoanions of fifth- and sixth-group transition metals. Although some of these polyanions have been known for over 100 years and are widely used in areas such as analytical chemistry and electron microscopy, the field as a whole has been inadequately treated in· currently available books.

The chemistry of polyoxometalate complexes, which can incorporate some 65 elements as heteroatoms in structures that range in size from 5-30 Å, bridges the chemistry of electrolyte solutions with that of extended lattices (semiconduction, heterogeneous catalysis, etc.). A brief historical review and a summary of experimental methods appropriate for the study of polyanions in solution is followed by a discussion of syntheses, structural principles, general properties and applications in analysis, biochemistry, catalysis, etc. Subsequent chapters treat isopolyanions, heteropolyanions, polyanions as ligands and cryptands, heteropoly blues, and organic- and organometallic derivatives. Possibilities for extending the field, and current views on bonding and mechanisms are discussed in the final chapter. Although the book forms part of a series in inorganic chemistry, the field of polyoxometalates deserves wider attention, for example, from organic chemists, especially those concerned wirh homogeneous and heterogeneous catalysis, and from biochemists, solid state- and materials scientists.

Springer-Verlag
Berlin
Heidelberg
New York

Volume 7
H. Rickert

Electrochemistry of Solids

An Introduction

1982. 95 figures, 23 tables. XII, 240 pages
ISBN 3-540-11116-6

The electrochemistry of solids is of great current interest to research and development. The technical applications include batteries with solid electrolytes, high-temperature fuel cells, sensors for measuring partial pressures of activities, display units and, more recently, the growing field of chemotronic components. The science and technology of solid-state electrolytes is sometimes called solid-state ionics, analogous to the field of solid-state electronics. Only basic knowledge of physical chemistry and thermodynamics is required to read this book with utility. The chapters can be read independently from one another.

Volume 6
D. L. Kepert

Inorganic Stereochemistry

1982. 206 figures, 45 tables. XII, 227 pages
ISBN 3-540-10716-9

An important recent advance concerns the stereochemistry of molecules containing ring systems, which are extremely important throughout chemistry. Such molecules may not have stereochemistries corresponding to any of the usual polyhedra, but are intermediate between two different idealized polyhedra. The precise location of a particular molecule along this continuous range of stereochemistries depends upon the geometric design of the ring system, which includes the number of atoms in ring and size of these atoms.

The simple techniques outlined in this work are the best way, and in most cases the only way, that such complicated structures with coordination numbers from four to twelve can be predicted.

Volume 5
T. Tominaga, E. Tachikawa

Modern Hot-Atom Chemistry and Its Applications

1981. 57 figures, 34 tables. VIII, 154 pages
ISBN 3-540-10715-0

Volume 4
Y. Saito

Inorganic Molecular Dissymmetry

1979. 107 figures, 28 tables. IX, 167 pages
ISBN 3-540-09176-9